T-Labs Series in Telecommunication Services

Series editors

Sebastian Möller, Berlin, Germany
Axel Küpper, Berlin, Germany
Alexander Raake, Berlin, Germany

More information about this series at http://www.springer.com/series/10013

Babak Naderi

Motivation of Workers on Microtask Crowdsourcing Platforms

 Springer

Babak Naderi
Quality and Usability Lab
Technische Universität Berlin
Berlin
Germany

Zugl.: Berlin, Technische Universität, Diss., 2017

ISSN 2192-2810   ISSN 2192-2829   (electronic)
T-Labs Series in Telecommunication Services
ISBN 978-3-319-89198-9   ISBN 978-3-319-72700-4   (eBook)
https://doi.org/10.1007/978-3-319-72700-4

Printed on acid-free paper

This Springer imprint is published by Springer Nature
The registered company is Springer International Publishing AG
The registered company address is: Gewerbestrasse 11, 6330 Cham, Switzerland

Acknowledgements

I would like to sincerely thank all people who supported me in doing my research in the last years and writing this dissertation. Over the past years, I have been fortunate to work together with brilliant colleagues at the Quality and Usability Lab of Technische Universität Berlin in an autonomy-supportive atmosphere.

Foremost, I would like to express my sincere gratitude to my principal supervisor Prof. Dr.-Ing. Sebastian Möller, who has continuously supported me and my work throughout the last years. Thank you for your endless patience and being a motivating and inspiring mentor for me. I also greatly appreciate that Prof. Dr. Tobias Hoßfeld and Dr. Sebastian Egger-Lampl agreed to be the co-examiner of my dissertation.

A considerable part of this thesis is a result of close collaboration with Dr.-Ing. Ina Wechsung. Thank you for guiding me toward the process of developing questionnaires, answering my countless questions about structural equation models, and statistical analyses. You have been a discerning and supportive colleague.

During the last years, I was closely collaborating with Dr.-Ing. Tim Polzehl and André Beyer in different projects. Thank you for being good and productive colleagues during the previous years, and wish you all the success with Crowdee! And thank you to my colleague Steven Schmidt for proofreading the first draft of this thesis. You have a keen eye for the details. Without that, I could not be on schedule. Thank you also to Irene Hube-Achter, Yasmin Hillebrenner, and Tobias Hirsch for the organizational support. We were so fortunate to have your support all the time.

I would like to thank all former and current colleagues at the Quality and Usability Lab, including Dr.-Ing. Dennis Guse, Dr.-Ing. Tilo Westermann, Dr.-Ing. Friedemann Köster, Dr.-Ing. Klaus-Peter Engelbrecht, Dr. Hamed Ketabdar, Dr.-Ing. Robert Walter, Dr.-Ing. Benjamin Bähr, Dr.-Ing. Jan-Niklas Voigt-Antons, Ph.D. Laura Fernández Gallardo, Dr. Benjamin Weiss, Dr.-Ing. Stefan Hillmann, Patrick Ehrenbrink, Gabriel Mittag, Maija Poikela, Saman Zadtootaghaj, Rafael Zequeira, Dr.-Ing. Justus Beyer, Dr.-Ing. Marc Halbrügge, and many more.

Nothing would have been possible without my family. Mama, you always support me in all steps in my life. Thank you deeply for always being behind me, listening and inspiring me to follow my dreams and realizing my potential. None of my works would be possible without your support. A very special thank you to my lifetime guide, my beloved and missed father who inspired me through life and has shaped who I am today. Bamdad, my brother, thank you for your support in pushing boundaries, and all of the good time we had together, mainly games that we played in childhood with your self-made imaginary characters! I believe they significantly exercised and boosted my imagination which is the main ingredient for success in our field. I am infinitely grateful for the support of you all over the years.

Finally and most importantly, I would like to thank my wonderful wife, Hywa and my lovely son, Nik. Hywa, thank you for your endless support, for enduring me burning the midnight oil, for the encouragement you gave me and standing beside me throughout this journey. I learned from you to never give up. Also, thank you Nik for giving me a deadline and listening to the dry runs of my speech with a full concentration!

Contents

Acronyms

ACBC   Adaptive Choice-Based Conjoint
ACR   Absolute Category Rating
AICC   Akaikes Information Criterion Corrected
API   Application Programming Interface
AVE   Average Variance Extracted
BIC   Bayesian Information Criterion
BPNT   Basic Psychological Needs Theory
BYO   Build Your Own
CA   Conjoint Analysis
CBC   Choice-Based Conjoint
CET   Cognitive Evaluation Theory
CFA   Confirmatory Factor Analysis
CFI   Comparative Fit index
CI   Confidence Interval
COT   Causality Orientation Theory
CR   Composite Reliability
CWMS   Crowdwork Motivation Scale
EFA   Exploratory Factor Analyses
GDP   Gross Domestic Product
HB   Hierarchical Bayes
HIT   Human Intelligence Task
HTML   Hypertext Markup Language
IMI   Intrinsic Motivation Inventory
IR   Item Reliability
IS   Inconsistency Score
KMO   Kaiser-Meyer-Olkin
LOT   Listening Only Tests
MOS   Mean Opinion Score
MTurk   Amazon Mechanical Turk
MVP   Minimum Viable Product

MW	Microworkers
OCR	Optical Character Recognition
OIT	Organismic Integration Theory
QoE	Quality of Experience
RAI	Relative Autonomy
RMSD	Root Mean Square Deviations
RMSEA	Root Mean Square Error of Approximation
RSME	Rating Scale Mental Effort
RTLX	Row Task Load Index
SDT	Self-Determination Theory
SGML	Standard Generalized Markup Language
SOS	Standard deviation of Opinion Scores
SRQ	Self-Regulation Questionnaires
TLX	Task Load Index
TO	Turkopticon
WEIMS	Work Extrinsic Intrinsic Motivation Scale

Abstract

Crowdsourcing microtask offers a fast, low-cost, and scalable approach to collect subjective data or solve problems that are still too complex for automatic processing and need human computation. Typical complex works can also be split up to some extent into simple tasks that can be performed through crowdsourcing microtasks. The range of possible microtasks is nearly countless, and potentially every computer-literate individual can be a crowdworker. Thus, studying the motivation of crowdworkers is crucial for the future of crowdsourcing microtasks to find out how to attract more people and reach a higher quality of outcomes.

In this book, first, a taxonomy for studying the motivation of crowdworkers is proposed including the potential influencing factors, different types of motivation, and possible consequences and outcomes related to the motivation. Next, the CWMS questionnaire, an instrument for measuring the underlying motivation of crowdworkers is developed. It considers different dimensions of motivation suggested by the self-determination theory of motivation which is a well-established and empirically validated psychological theory used in various domains. This instrument can be used to study the effect of platform and user characteristics on the general motivation of crowdworkers. Later, the task-specific motivation of crowdworkers is studied in detail: Influencing factors are investigated, subjective methods for measuring them are evaluated, a model for predicting worker's decision on taking a task is proposed, the relative importance of different factors for two populations of crowdworkers is studied, and finally, a model for predicting the expected workload (as one of the major influencing factors) given the task design is proposed. Last but not least, the effect of worker motivation and task design on the performance of crowdworkers is analyzed.

Results show that workers decide to take a task mostly based on its payout, expected workload, and interestingness, and in some cases how recently it was published and how many instances are available. Fairness and generosity of employers are also important for them. Their general motivation of crowdworking insignificantly influences the degree of importance of those factors. Users' characteristics such as skills, experiences, and preferences influence the perceived task's interestingness and estimation of workload. Workers with intrinsic or internalized extrinsic motivation participate more and present more reliable answers when working on specific tasks. However, the degree of external and identified motivation relates to a worker's overall (long-term) reliability score. As this score is widely used by employers to select workers who can perform their job, serious workers, despite their underlying motivation, try to keep it high. Furthermore, applying reliability check methods, and acknowledging workers' reliable answers through the task design are recommended as results show that they encourage workers to provide high-quality answers.

Zusammenfassung

Crowdsourcing Micro-Task bietet einen schnellen, kostengünstigen und skalier-baren Ansatz um subjektive Daten zu sammeln oder Probleme zu lösen, die immernoch zu komplex für eine automatische Verarbeitung sind und daher nach einer Bearbeitung durch Menschen verlangen. Typische komplexe Arbeitspakete können zu einem gewissen Maße in einfachere Aufgaben aufgeteilt werden, die mittels Crowdsourcing Micro-Tasks bearbeitet werden können. Die Einsatzmöglichkeiten von Micro-Tasks sind fast zahllos und potenziell kann jede Person, die sich mit Computern auskennt, ein Crowdarbeiter sein. Daher ist es für die Zukunft von Crowdsourcing Micro-Task entscheidend, die Motivation von Crowdworkern zu untersuchen, um mehr Personen anzusprechen und eine höhere Antwortqualität zu erzielen.

In dieser Dissertation wird zuerst eine Taxonomie zum Studium der Motivation von Crowdarbeitern vorgestellt, die potenzielle Einflussfaktoren, verschiedene Motivationstypen und mögliche Konsequenzen und Resultate bezogen auf die Motivation beinhaltet. Danach wird der CWMS-Fragebogen, ein Instrument zum Messen der zugrundeliegenden Motivation von Crowdarbeitern, entwickelt. Es betrachtet verschiedene Dimensionen der Motivation, die durch die Selbstbestimmungstheorie der Motivation, die eine gut etablierte und empirisch validierte psychologische Theorie ist und in verschiedenen Bereichen verwendet wird, vorgeschlagen werden. Dieses Instrument kann verwendet werden, um die Wirkung von Plattform- und Benutzereigenschaften auf die allgemeine Motivation der Crowdarbeiter zu untersuchen. Dann wird die aufgabenbezogene Motivation von Crowdarbeitern im Detail betrachtet: Einflussfaktoren werden untersucht, subjektive Methoden zu ihrer Messung werden ausgewertet, ein Modell zur Vorhersage der Entscheidung der Arbeiter für eine bestimmte Aufgabe wird vorgeschlagen, der relative Einfluss verschiedener Faktoren für zwei Populationen von Crowdarbeitern wird untersucht und schließlich wird ein Modell für die Vorhersage der erwarteten Arbeitsbelastung (als einer der wichtigsten Einflussfaktoren) verbunden mit dem Aufgabendesign vorgeschlagen. Nicht zuletzt wird die Wirkung von der Arbeiter-Motivation und dem Task-Design auf die Leistung von Crowdarbeitern analysiert.

Die Ergebnisse zeigen, dass die Entscheidung der Arbeiter eine Aufgabe zu bearbeiten, hauptsächlich auf der Auszahlungshöhe, der erwarteten Arbeitsbelastung und darauf beruht, ob die Aufgabe als interessant wahrgenommen wird und in einigen Fällen, vor wie langer Zeit sie veröffentlicht wurde und wie viele Aufgaben verfügbar sind. Fairness und Großzügigkeit der Arbeitgeber sind hierbei für die Motivation der Arbeiter auch wichtig. Ihre allgemeine Motivation für Crowdworking beeinflusst die Bedeutung dieser Faktoren nicht signifikant. Die Eigenschaften der Crowdarbeiter, wie ihre Fähigkeiten, Erfahrungen und Präferenzen, beeinflussen die wahrgenommene Interessantheit von Aufgaben und die Einschätzung der Arbeitsbelastung. Arbeiter mit intrinsischer oder internal-isierter extrinsischer Motivation beteiligen sich stärker und liefern zuverlässigere Antworten bei der Arbeit an bestimmten Aufgaben. Jedoch hängt der Grad der externen und identifizierten Motivation mit der Gesamtbewertung für die (Langzeit-)Zuverlässigkeit ("Reliability") eines Arbeiters zusammen. Da diese Bewertung von den Arbeitgebern verwendet wird, um Arbeiter auszuwählen, die ihre Aufgaben bearbeiten können, versuchen ihn ernsthafte Crowdarbeiter ungeachtet ihrer zugrundeliegenden Motivation hoch zu halten.

Darüber hinaus wird die Anwendung von Methoden der Zuverlässigkeitsprüfung und die Bestätigung von validen Antworten der Arbeitnehmer durch die Aufgabengestaltung empfohlen, da die Ergebnisse zeigen, dass diese die Crowdarbeiter dazu ermutigen, qualitativ hochwertige Antworten zu geben.

Chapter 1
Introduction

The term *crowdsourcing*, introduced in 2006 by combining *crowd* and *outsourcing*, refers to "the act of taking a job traditionally performed by a designated agent (usually an employee) and outsourcing it to an undefined, generally large group of people in the form of an open call" [41, 49]. In essence, crowdsourcing utilizes the potential ability of a large network of people, who are connected through the Internet, to accomplish a particular task [34]. Over the time, numerous systems have appeared on the Internet to solve a broad range of problems, by applying different collaboration methods under the umbrella of crowdsourcing [22]. Famous examples from different domains include Wikipedia, Linux, TripAdvisor, Stack Overflow, and Amazon Mechanical Turk[1] (MTurk).

Different attempts were made to classify the crowdsourcing systems by specifying who are the people behind the crowd, what they do, why and how they do it in various levels of detail [22, 26, 70, 118]. Besides the non-profit systems like Wikipedia and Stack Overflow, the commercial-crowdsourcing platforms in which the crowd are paid in exchange for their service, gained more attention and expanded in various dimensions [41]. These platforms act usually as a mediator and provide tasks given by employers to the crowd either in the form of contest or open-call [41]. In some platforms, a specialized group of workers participate in a contest for example to design a logo (e.g. 99design) or to solve a challenging scientific problem (e.g. InnoCentive) which is typically reimbursed by a five-figure reward. Other platforms, called *crowdsourcing microtask*, provide a broad range of small tasks in an open-call, to a large and undefined crowd which are usually reimbursed by a small monetary reward per each piece of work they perform. This dissertation primarily aims to understand the crowd that works on the crowdsourcing microtask platforms and investigate their motivation. The motivation is a crucial aspect for the future of crowdworking as it influences the performance and participation of crowdworkers [63].

---

[1]https://www.mturk.com.

© Springer International Publishing AG 2018
B. Naderi, *Motivation of Workers on Microtask Crowdsourcing Platforms*, T-Labs
Series in Telecommunication Services, https://doi.org/10.1007/978-3-319-72700-4_1

1.1 Crowdsourcing Microtask

Crowdsourcing microtask provides a remarkable opportunity for academic and industry sectors by offering a high scale, on demand and low-cost pool of geographically
distributed workforce for completing complex tasks that can be divided into a set of
short and simple online tasks like annotations and data collection [44, 60, 63, 81].
Typically, *crowdworkers* are paid on a piecework basis. *Tasks* are short, often for-
mated in a form-like webpage and are as simple as a computer-literate worker should
be able to perform one in couple of minutes. Tasks are offered using a crowdsourcing
platform which manages the relationship between crowdworkers and *job provider*. A
job provider creates a *job* in the crowdsourcing platform which is a collection of tasks
that can be done by one or many crowdworkers in exchange for monetary rewards.
In the following, crowdsourcing microtask platforms and tasks are described.

1.1.1 Platforms

The platforms are maintaining a dedicated crowd of workers and providing required
infrastructure like a pool of tasks for workers, payment mechanisms, and in some
cases additional services like quality control or worker selection mechanisms [44].
MTurk is the best known, most used and most researched platform within the academic literature and in public [72]. On average, 708,000 tasks were daily available
on MTurk in the first quarter of the year 2017.[2]

The crowdsourcing platforms may provide a web access (e.g. MTurk, and
Microworkers[3]) or a native mobile application (e.g. Clickworker,[4] and Crowdee[5])
for crowdworkers. In addition, some platforms not only entertain their own crowd
but also connect to one or more other crowdsourcing platforms at the API level. As a
result, they may automatically push specific jobs, created on their platform, to one or
more other platforms and deliver the collected answers to the job provider afterward
(e.g. CrowdFlower[6]). As a result, they reach a widely spread crowd of workers on-
demand without creation and maintenance fees rather by paying per-usage hosting
costs to the other platforms.

Figure 1.1 illustrates a simplified crowdworking process following the MTurk
and the Crowdee platform structure (excluding the payments). First, the job provider
creates a job on the platform which may contain external content to be assessed by
the workers (process A in Fig. 1.1). Usually, platforms provide different templates
for jobs (e.g. survey job, moderation of an image, or sentiment analysis project), and
also permit job providers to create a custom designed job using raw HTML code. A

---

[2]Collected statistics by MTurk Tracker [21] were aggregated.
[3]https://microworkers.com.
[4]https://www.clickworker.com.
[5]https://www.crowdee.de.
[6]https://www.crowdflower.com/.

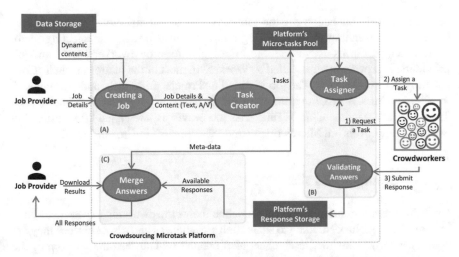

Fig. 1.1 General workflow of the crowdsourcing microtask platforms. **a** The job providers create a job with(out) dynamic content. **b** Crowdworkers visit the platform and ask for a task from one job, for which later they will submit an answer. **c** The job provider can download the latest list of answers at any time

job acts as a customized template. The job provider also specifies properties of the job (e.g. reward per assignment, and the number of repetitions requested per task), and worker's requirements (e.g. location of crowdworker). During the task creation process, one or more tasks will be created for the job depending on the number of dynamic contents and the requested repetitions. For instance, a survey job with 100 repetitions leads to 100 tasks and an image labeling job with 100 images and three repetitions will leads to 300 tasks. Tasks will be added to the pool of available microtasks and will be ready to be performed by the crowdworkers. Furthermore, the platform holds corresponding funds (i.e. the sum of rewards expected to be paid to the crowdworkers, and platforms fee) for liability.

Meanwhile, the crowdworkers browse the list of available jobs in the platform. By selecting a particular job, a crowdworker requests a task from that job to be reserved and assigned to him/her (process B in Fig. 1.1). After performing the task, the crowdworker will submit the corresponding answer. The submitted answers are stored in the platform's repository of answers.

Last but not least, job providers can inquire the list of submitted answers (process C in Fig. 1.1). Typically, the job provider has a limited time to proof the submitted answers and accept or reject them. Otherwise, the platform will accept the submitted answer automatically. By accepting the answer, the corresponding crowdworker will be paid, and by rejecting it, the crowdworker get a notification and the corresponding task will return to the pool of available tasks to be taken by other workers.

Platforms may offer additional features, including different job templates, pre-checking of the job by experts (e.g. CrowdFlower), rewards estimation (e.g. Microworkers), and built-in quality control (e.g. CrowdFlower). Typically, platforms provide an API connection for the job providers to automate the process of creating a job, appending tasks, inquiring answers and approving or rejecting them. In addition, platforms use different terminologies for crowdworkers, tasks and job providers. A crowdworker is called a "Worker" in most of the platforms but a "Contributor" in CrowdFlower. The job is called a "Project" in MTurk, and "Campaign" in Microworkers. A task (an instance of job) is called a "HIT" (Human Intelligence Task) or an "Assignment" in MTurk. The job providers, are called "Requesters" in MTurk, "Employers" in Microworkers and "Customers" in CrowdFlower.

1.1.1.1 Crowdworking Tools

Crowdworkers use different tools to enhance their crowdworking experience. They use forums (e.g. Turk Nation[7]) to share their experiences and discuss issues they are dealing with. Meanwhile, researchers also created tools to use the wisdom of the crowd to support themselves.

Turkopticon[8] (TO) [56] is an activist system that allows crowdworkers to evaluate their relationships with requesters using publicly visible ratings. Using the TO extensions, for Internet browsers like Firefox or Chrome, crowdworkers can see aggregated ratings of each requester on MTurk listings. The TO ratings are highly used by crowdworkers and are also integrated into nearly all scripts they use. Turkopticon provides five-points ratings for the four dimensions [56]: Communicativity, Generosity, Fairness and Promptness. Further details are given in Sect. 5.4.

Turkmotion,[9] developed by the author of this book, followed the same procedure and let workers rate how enjoyable a HIT is and how adequately it is rewarded for each HIT in MTurk. By installing the Turkmotion extensions for Internet browsers, crowdworkers can rate each HIT or see aggregated ratings of each HIT on MTurk listings.[10]

1.1.2 Tasks

Crowdsourcing microtask can be used for various purposes. Examples are supporting Artificial Intelligence with image tagging, audio/video transcribing, and subjective quality assessment. Gadiraju et al. [30] suggested six categories of microtasks:

- Information Finding: Searching Web for a certain information.
- Verification and Validation: Verifying certain information or confirming the validity of a piece of information.

---

[7]http://turkernation.com/.
[8]http://turkopticon.ucsd.edu/.
[9]http://turkmotion.com/.
[10]Instead of starting hype, the system is not utilized by the crowdworkers.

- Interpretation and Analysis: Interpreting Web content. For example, categorizing product pictures in a predefined set of categories.
- Content Creation: Generating new content. For example, summarizing a document, transcribing an audio document or describing a picture.
- Surveys: Answering a set of questions related to a certain topic.
- Content Access: Accessing some Web content. For example, watching an online video.

The range of possible microtasks is nearly innumerous. It is bounded to what is technically possible (transferable via the Internet in a form-like format), and the skills and ingenuity of those designing and carrying out the work [63, 72]. Currently, there are thousands of individuals behind the crowdsourcing microtask platforms but nearly, every computer-literate individual is a potential crowdworker. The skills, ability, and diversity of crowdworkers shape the functionalities and define the scope of services that crowdsourcing microtask can provide. Considering the diversity of crowdworkers and them being the dominant aspect of process, which shape the provided services through crowdsourcing microtask, it is crucial for the *future of crowdworking* [63] to understand their motivation, factors affecting it, and outcomes influenced by it.

1.2 Scope and Research Questions

This dissertation addresses essential questions regarding crowdworkers:

- Who are crowdworkers?
- What are the factors that can influence the motivation of crowdworkers?
- How to measure the general motivation of crowdworkers? Are there different types of motivation? If yes, what is their influence?
- Is there a task-specific motivation? What factors are influencing it (including the general motivation)? How to measure them?
- Is workers' task selection behavior predictable?
- How does motivation influence the performance of workers? How to check the reliability of answers collected through crowdworking?

This book organized as follows. Chapter 2 presents theories that are central to understand crowdworking motivation and answers to the above questions on what factors of crowdsourcing microtask are connected to the motivation of workers. A taxonomy of crowdworking motivation is presented.

Chapter 3 answers the question of who crowdworkers are, particularly regarding demographic information on different platforms.

Chapter 4 deals with the question of how the general underlying motivation of crowdworkers can be measured. As a result of this chapter, a new questionnaire, the CWMS, is developed.

Chapter 5 focuses on the task-specific motivation of crowdworkers and answers the above questions of what factors are influencing this motivation, how they can be measured, what their relative importance is, and whether the task selection of crowdworkers is predictable. In addition, expected workload, as a major factor, is further analyzed and models for predicting it are provided.

Chapter 6 focuses on aspects influencing workers' performance, namely their motivation and the task design.

Finally, Chap. 7 summarizes findings of the previous chapters and closes the dissertation with an outlook on future work. Note that, methods used to check the reliability of answers and the data screening process are described in various places as they are crucial parts of each crowdsourcing study.

Publications used in this book are listed below:

- Naderi, B., Polzehl, T., Beyer, A., Pilz, T., Möller, S.: Crowdee: Mobile crowdsourcing microtask platform–for celebrating the diversity of languages. In: Proceedings 15th Annual Conference of the International Speech Communication Association (Interspeech 2014). IEEE (2014).
- Naderi, B., Wechsung, I., Polzehl, T., Möller, S.: Development and validation of extrinsic motivation scale for crowdsourcing micro-task platforms. In: Proceedings of CrowdMM 14, pp. 3136. ACM (2014).
- Naderi, B., Wechsung, I., Möller, S.: Crowdsourcing work motivation scale: development and validation for crowdsourcing micro-task platforms. Submitted to Behaviour and Information Technology.
- Naderi, B., Wechsung, I., Möller, S.: Effect of being observed on the reliability of responses in crowdsourcing micro-task platforms. In: Quality of Multimedia Experience (QoMEX), pp. 12. IEEE (2015).
- Naderi, B., Polzehl, T., Wechsung, I., Kster, F., Möller, S.: Effect of trapping questions on the reliability of speech quality judgments in a crowdsourcing paradigm. In: 16th Annual Conference of the International Speech Communication Association (Interspeech 2015). ISCA, pp. 2799–2803 (2015).
- Martin, D., Carpendale, Gupta, N., Hoßfeld, T., Naderi, B., Redi, J., Siahaan, E., Wechsung, I.: Understanding the crowd: ethical and practical matters in the academic use of crowdsourcing. In: Evaluation in the Crowd. Crowdsourcing and Human-Centered Experiments. Springer, Cham (2017).

Chapter 2
Theoretical Background on Motivation

In this chapter, a taxonomy of motivation of crowdworkers is presented which is empirically evaluated during this dissertation. The taxonomy is base on the Self-Determination Theory (SDT) [19, 100] of motivation and its framework for studying motivation. Relevant findings from SDT in other domains including the work domain are adapted to crowdworking to develop the taxonomy. Later, the Herzberg two-factor theory of motivation is briefly introduced, and a summary of relevant research in the domain of crowdworking is given.

The taxonomy is illustrated in Fig. 2.1. It consists of three layers: (1) Different factors which influence motivation, (2) different types of motivation from the perspective of the SDT, and (3) relevant outcomes which might be influenced by motivation. According to the SDT the quantity of a person's behavior is related to the amount of his/her motivation, but the quality of a person's behavior is associated with the type of motivation which also directly influences a person's wellbeing [19]. The motivation layer was partly addressed in [84, 85].

2.1 Taxonomy

2.1.1 Motivation

Motivation is defined as "a reason or reasons for acting or behaving in a particular way" [107]. Here we distinguish between *general underlying motivation* of working in a crowdsourcing platform and the *task-specific motivation* which refers to why one choses a particular task to perform given various options usually available in the platform. The general crowdworking motivation is hypothesized to influence the task-specific motivation.

© Springer International Publishing AG 2018   7
B. Naderi, *Motivation of Workers on Microtask Crowdsourcing Platforms*, T-Labs
Series in Telecommunication Services, https://doi.org/10.1007/978-3-319-72700-4_2

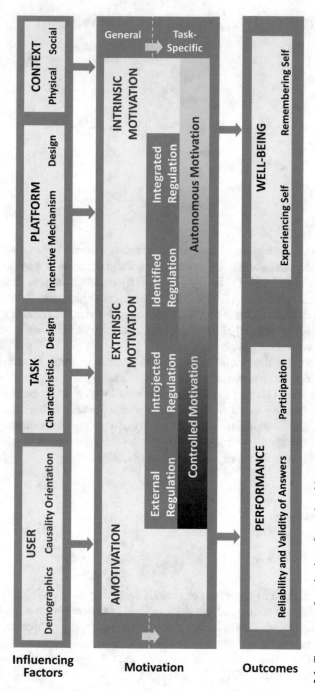

Fig. 2.1 Taxonomy of motivation of crowdworking

Theories from the field of work motivation often differentiate among two distinct types of motivation by their source of origin, i.e. *intrinsic* and *extrinsic* motivation. Intrinsic motivation refers to activities where the origin of the motivation is driven by an internal force, which consequently returns spontaneous satisfaction to oneself from performing the activity itself [32]. In contrast, extrinsic motivation leads to the satisfaction that is not rooted in the activity itself, rather in separable consequences of the activity, such as (monetary) rewards or the avoidance of punishment [32].

Despite the fact that intrinsic motivation has a positive effect on performance, in reality, many tasks in the work domain (including the majority of jobs on paid crowdsourcing platforms) are not primarily composed to induce intrinsic motivation. This is also true for many other circumstances, as Ryan and Deci [100] stated:

> Indeed, much of what people do is not, strictly speaking, intrinsically motivated, especially after early childhood when the freedom to be intrinsically motivated is increasingly curtailed by social pressures to do activities that are not interesting and to assume a variety of new responsibilities.

In order to study the relationship between motivation, driving forces, and outcome, the SDT can be applied.

2.1.1.1 Self-determination Theory (SDT) of Motivation

In essence, SDT distinguishes between *autonomous* motivation and *controlled* motivation. Accordingly, the theory recommends that behaviors can be classified based on the degree of autonomous or controlled motivation. Autonomy involves "acting with a sense of volition and having the experience of choice", whereas control refers to "acting with a sense of pressure, a sense of having to engage in the actions" [32]. Thus, intrinsic motivation is an example of autonomous motivation, whereas extrinsic reward belongs to controlled motivation. Moreover, the SDT addresses the process of *internalization* of extrinsically motivated activities [32]. Internalization refers to "the process through which an individual acquires an attitude, belief, or behavioral regulation and progressively transforms it into a personal value, goal, or organization" [19]. Despite the primary classification of motivation to intrinsic and extrinsic, which only consider the activity, the SDT considers both, the activity and the self.

The SDT introduces six different types of motivations for an activity, considering its perceived locus of control. It includes amotivation (the absence of motivation), a spectrum of the extrinsic motivation, and the intrinsic motivation. Figure 2.2 illustrates the continuum of motivation as proposed by the SDT. Each of them is briefly explained in the following:

Intrinsic Motivation. It is defined by an inner force being the cause of the motivation and leads to spontaneous satisfaction. *Interest and Enjoyment* are considered as direct measurements of intrinsic motivation. It is defined as the activity itself being perceived interesting or enjoyable [75].

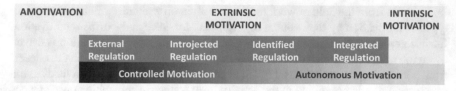

Fig. 2.2 The SDT continuum of motivation, after [32]

External Regulation. It refers to a situation in which an activity being initiated and maintained by contingencies external to the person. As a result of external regulation, one acts with the intention of achieving a desirable consequence or avoiding an undesirable one [32]. Performing a task to earn (monetary) rewards is an example of external regulation in crowdworking.

Other types of extrinsic motivation occur when the regulation of the activity and the values linked to it have been internalized. They have a different degree of internalization:

Introjected Regulation. It refers to regulatory motivation that its locus of causality is somewhat external. It has been taken in by the person but has not been admitted as his or her own [32]. Examples are contingent self-esteem and ego involvement, which cause pressure and trigger a particular behavior to feel valuable or to support the ego. Performing a task to exhibit the own ability is an example of introjected regulation in the crowdworking domain.

Identified Regulation. It describes situations in which the behavior is somewhat internalized. The activity is identified as a self-selected goal and valuable; greater freedom and volition is experienced. Thus, the behavior is perceived to have an internal locus of causality. Examples are crowdworkers, who are identifying themselves with the values of the job, and goals of the job and job provider.

Integrated Regulation. It refers to the fullest type of internalization. The activity, its goals, and values are largely internalized and are perceived as an integral part of the self. Integrated regulation shares some characteristics with intrinsic motivation. An example is a job posted by a particular community (e.g. a political party) and a worker performing the job because he/she felt being part of the community and accepted their goals and values as his/her.

The intrinsic, integrated and identified regulations are joined to create the autonomous motivation whereas external and introjected regulations sum up to the controlled motivation.

Amotivation. It refers to a lack of motivation and intention. Both autonomous and controlled motivations are intentional; Thus, both are in contrast to amotivation.

The very first attempts made for developing an instrument to measure the motivation of crowdworkers based on the SDT were reported in [84, 85]. Measurement of the general underlying motivation of crowdworkers is addressed in Chap. 4, and task-specific motivation is addressed in Chap. 5.

2.1.2 Influencing Factors

It is assumed that motivational influencing factors have a direct and indirect effect on the general underlying and task-specific motivation of crowdworkers. They include the characteristics of the user (i.e. crowdworkers), the given task, platform, and context (cf. Fig. 2.3). The SDT includes six sub-theories which were tested and validated empirically in different domains. Here they are used to describe the potential influencing factors in the domain of crowdworking.

User. All characteristics of the user which carry an influence on his/her motivation and performance should be considered. In most of the studies, demographic data such as gender, and age are usually asked and reported. Age may relate to a decrease of the cognitive ability [114, 117] and perception of another level of difficulty and enjoyment in the domain of crowdsourcing. Educational attainment relates to skills and experience which may influence the perceived workload associated with a task and related motivation. Furthermore, the importance of task-related factors influencing motivation may differ based on their country of living. People living in the Western countries may consider digital workload more important than one living in a developing country. Finally, workers' household size and income influence their goal and how serious they take crowdworking. Another aspect is the causality orientation of the user which is described by the Causality Orientation Theory (COT), a sub-theory of SDT. It describes individual differences in motivation orientation i.e. autonomously or controlled orientation. The autonomously-oriented individuals tend to organize their behaviors based on their intrinsic interest whereas control-oriented individuals follow deadlines and rewards [65]. The theory hypothesizes that motives corresponding to individual orientation lead to a better outcome. Demographic and causality orientation factors mostly influence the general underlying motivation of workers.

Task. The task attributes such as enjoyment, reward, workload, and value may influence the task-specific motivation. The task design like long unclear instructions, or applying reliability check methods can influence both the motivation and performance of workers. In addition, another sub-theory of SDT, the Cognitive Evaluation Theory (CET) [19] addresses the influence of contextual elements on intrinsic motivation. CET warns that extrinsic motivators such as deadlines and tangible rewards can cause a significant decrease or even a collapse of intrinsic motivation in particular settings.

Platform. The platform design and incentive mechanisms exert influence over the general motivation and (indirectly) on the performance of crowdworkers through sat-

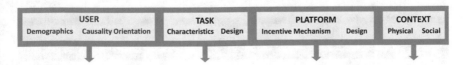

Fig. 2.3 Factors influencing the general and task-specific motivation

isfying the basic psychological needs (i.e. autonomy, competence, and relatedness). The Basic Psychological Needs Theory (BPNT) [20] and Organismic Integration Theory (OIT) [100] postulate that, when the basic psychological needs are satisfied concerning a behavior, the value and regulation of the activity will be internalized. Those needs are defined as "universal necessities, as the nutriments that are essential for optimal human development and integrity" [32]. Baard et al. [3] have shown the strong positive influence of perceived autonomy support work climate (i.e. environment supports intrinsic motivation and facilitates internalization of extrinsic motivation) on need satisfaction, performance, and well-being. The influence of platforms is not investigated in this dissertation but included in the taxonomy for the sake of completeness.

Context. The physical and social context [94] may influence the general motivation and also the performance of crowdworkers. The physical context refers to the properties of location and space in which crowdworkers are performing their task. The physical context may change during the time in case of desktop crowdworking (e.g. turning on the television in the living room) and by movement of the worker in the case of mobile crowdworking (e.g. using a public transport system). Environmental distraction can influence the performance of crowdworkers on most of the tasks including the speech quality assessment in which some degradated stimuli may judge better or worse in respect to their quality than how they would have been judged in the laboratory environment [77]. The social context mostly influences the task-specific motivation. Workers may consider reviews written about a job provider, by other workers, when considering to accept the job. The influence of physical context is not investigated in this dissertation but included in the taxonomy for the sake of completeness.

2.1.3 Outcomes

The type and the amount of motivation for performing an activity influences the *performance* and subjective *well-being* (cf. Fig. 2.4).

Performance. The *reliability and validity of responses* and the amount of *participations* are considered to be two measures which indicate quality and quantity of job-related outcomes. The reliability refers to the extent to which the responses are consistent over the time. The validity refers to how truthful the responses are and if the response truly is what it was intended to be. To measure the validity of

PERFORMANCE		WELL-BEING	
Reliability and Validity of Answers	Participation	Experiencing Self	Remembering Self

Fig. 2.4 Outcomes hypothesized to be influenced by the motivation

responses a ground truth or alike is needed. For instance, the validity of a response to a quality assessment task in crowdsourcing can be examined by comparing results from a crowdsourcing test and results from laboratory tests assuming the later to be close to the ground truth. In case that the ground truth is unknown, the reliability of response can be examined by introducing new questions in which their answers are known. Different methods in this regards are presented in Sect. 4.2.1.1. Task characteristics like type (e.g. simple or complex) and interestingness determine the effect of motivation on the performance. Laboratory experiments and field studies in several domains have shown that for complex tasks (involving flexibility and creativity) autonomous motivation yields more effective performance whereas controlled motivation leads to either no difference or a small advantage in simple tasks (e.g. tedious application of an algorithm) [84] (for an overview see [32]). For interesting tasks, the intrinsic motivation leads to a better performance whereas autonomous extrinsic motivation results in greater performance on non-interesting tasks, which need discipline or determination [32]. An example is voting behavior: Koestner et al. [64] discovered when people are intrinsically interested in the issues they become well informed, but it is more likely that they make an effort to go out and vote when they are motivated by the importance of the issue for themselves (i.e. autonomous extrinsic motivation).

Well-being. Autonomous motivation is associated with greater job satisfaction and well-being [32]. In the domain of well-being, Kahneman [57] addressed a conflict between *Experiencing Self* and *Remembering Self*. The experiencing-self answers to a question of how one feels at a certain moment; in contrast, the remembering-self keeps an overall score. It reflects the person's global evaluation of an entire period of time in the past and influences one's future decision-making. The experiencing-self and the remembering-self are satisfied differently with respect to environmental characteristics. The effect on the worker's well-being is not investigated in this dissertation but included in the taxonomy for the sake of completeness.

2.2 Other Motivational Theories

Two other relevant motivational theories are Herzberg's Two Factor Theory [39] and Warr's Vitamin Model [113]. The Two Factor Theory [39] distinguishes between *hygiene factors* and *motivators*. Hygiene factors or dissatisfiers are factors whose presence do not influence job satisfaction, but their absence will result in dissatisfaction. They associate with one's relationship to the context or environment in which the job is performed [48]. Examples are efficient procedures, working conditions, and salary [48]. In contrast, the absence of motivators or satisfiers does not lead to dissatisfaction, but their presence will facilitate satisfaction and motivation which will eventually result in better performance [39]. Examples include job content factors such as achievement and acknowledgment. These dimensions are not opposite ends of the same continuum, but rather they represent two distinct continuums [48].

Warr introduced the Vitamin Model which indicates that mental health, including affective well-being, is influenced by job and environmental characteristics in a non-linear way [113]. Some characteristics (mostly extrinsic job features) are important for mental health up to, but not beyond, a certain level (e.g. salary and task significance), others characteristics (mostly intrinsic job features) will even become harmful in very large quantities (e.g. job autonomy and social support).

Here a contradiction between Herzberg's theory and the Vitamin Model exists as the later considers an inverted U-shaped curvilinear relation between job characteristics like job demands, and autonomy and the employee well-being. Both of the abovementioned theories consider the effect of influencing factors on the worker's satisfaction and mental health and provide less evidence on the relation to their performance [23]. In contrast, SDT is a complete framework for studying the motivation which addresses the three layers of influencing factors, motivation, and outcomes as well as their interaction. It considers different types of motivation, their effect on both performance and well-being as outcome, and details how to design the work environment to facilitate the internalization of worker's primary motivation also considering personal differences. It has been applied to various domains including education, sport, and work and supported by empirical data. Consequently, SDT was chosen to be followed as the main theory in this book without neglecting the abovementioned theories.

2.3   Motivation in Crowdsourcing

Crowdworkers have a wide range of motives and experiences [61, 63]. Although a significant portion of their motivation is derived from monetary rewards, at least parts of it have its source in other aspects such as having fun or killing time [2, 9, 55].

Kaufmann et al. [58] proposed a model for motivation in crowdworking differentiating between intrinsic and extrinsic motivation. Their model contains enjoyment (i.e. task autonomy, task identity, and pass-time) and community-based (i.e. community identification and social contact) motivation as part of the intrinsic motivation. On the other side, payoffs (immediate and delayed) and social motivators are components of the extrinsic motivation. According to the model by Kaufmann et al. community-based motivation, which belongs to intrinsic motivation, is created when a worker is drove by social contact or by identification with the community. Social motivation, which is unlike community-based motivation referring to extrinsic motivation, is created by the social motivators that are originating from outside the platform community. Their results showed that extrinsic motivation strongly influences the time workers spent on the platform. The payoff was the most dominant motivator, followed by the intrinsic motivation aspects, especially enjoyment-based motivation like task autonomy and skill variety. Note, that assuming the community-based motivation as a component of the intrinsic motivation contradicts with the definition by

Gagn and Deci [32] described above because the satisfaction is not originated in the activity itself but rather in something external (e.g. community identification).

Rogstadius et al. [96] studied the effect of intrinsic and extrinsic motivation in paid crowdworking. They found that increasing the level of extrinsic motivation (i.e. payout) increases the workers' willingness to accept a task which leads to faster completion time but does not influence the accuracy of the outcome. However, decreasing the intrinsic motivations decreases the output quality. Likewise, their conception of intrinsic motivation does not match the definition by Gagne and Deci [32] as Rogstadius et al. assumed that "framing a task as helping others leads to increase of intrinsic motivation" [96]. Therefore, the resulting satisfaction is not gained from the activity itself but rather from the separable outcome, i.e. the feeling of helping others.

Others also studied the direct effect of task design, financial incentives and rewards schemes on the quality and quantity of outcomes. Chandlera and Kapelner [12] reported that framing a task as a meaningful and valuable activity increased the quantity of output (with an insignificant change in quality). Furthermore, the quality of outcome decreased (with no change in the quantity) when no context was given to workers, instead they were told that responses would be discarded after the study. Finnerty et al. [29] reported the influence of rewards scheme and task design on the quality of output. They found that a dynamic incentive system leads to better results when the amount of rewards is calculated based on both time and accuracy of the response. Meanwhile, they showed that a clearer and simpler task design, which demands less cognitive attention of worker, leads a to more accurate result. Mason and Watts [74] also reported that increasing the financial incentives in crowdworking increase the quantity, but not the quality, of work.

For the crowdsourcing contest, Zhao and Zhu [119] investigated the influence of motivation type on the amount of effort individuals spent. Following the SDT concepts of motivation, they discovered that external (i.e. rewards), introjected (i.e. ego-enhancement), and intrinsic motivation types are significantly and positively associated with efforts a participant spent on the contest. Within voluntary crowdworking, Borst [6] found out that intrinsic motives, such as fun and learning play a major role. She figured out that the *peer recognition* is the main extrinsic motive for the online-volunteer participants. Finally, Redi and Povoa [91] compared outputs from volunteer workers and paid crowdworkers. They infer that, compared to the paid workers, the volunteers are less likely to finish all tasks but when they complete tasks, their responses were more reliable than the answers of the paid workers.

However, none of the abovementioned studies measured the motivation using a validated instrument. The effect of job characteristics was separately studied, but no study (to the best of author's knowledge) specified the relative importance of task characteristics on the task-specific motivation. The effect of general motivation on the task-specific motivation is not investigated and the influence of motivation type (as understood by SDT) on performance is not empirically tested.

2.4   Chapter Summary

This chapter describes a taxonomy for motivation in crowdworking. The taxonomy is created based on the theoretical assumptions from the SDT which were validated in different domains including the work domain. In the central layer, the motivation of workers is presented as described by SDT, i.e. intrinsic motivation, a spectrum of extrinsic motivation and amotivation. The taxonomy distinguishes between the general underlying motivation of crowdworking and the task-specific motivation. The later refers to the individual preference on selecting a task to perform.

On the top layer, hypothesized influencing factors are listed. User's characteristics and motivation orientation, platforms design and its incentive mechanism, all are assumed to affect the general crowdworking motivation. The task characteristics and design are expected to mainly influence the task-specific motivation.

The abovementioned factors also influence the outcomes namely job performance and well-being of workers directly or indirectly through the motivation type. The performance of crowdworkers is represented by the reliability of their responses (general or task-specific) and the amount of participation.

The taxonomy provides an overview of aspects which will be studied in next chapters.

Chapter 3
Who are the Crowdworkers?

Crowdworkers are the key component and the main drive of the crowdsourcing micro-task platforms. Their skills, ability, and diversity shape the functionalities and define the scope of services that can be provided by platforms. Depending on the platform, crowdworkers exhibit a different degree of diversity in gender, education, the country of origin, and socioeconomical background. Understanding the demographics of crowdworkers is the first step in understanding their motivation. Demographics like education level can influence job's satisfaction and motivation [111]. Other traits such as gender and location are related to differences in the workers' performance and data quality [59].

There are several studies looking for understanding the population of crowd-workers. Survey-based demographic studies on MTurk [54, 55, 97] show that the majority of the workers population in 2010 were U.S ($\approx$50–60%) and Indian ($\approx$30–40%) based workers. The U.S workers were apparently more female ($\approx$65%) and 30+ years old on average. Indian workers were more male ($\approx$70%) than female and on average slightly younger, with 26–28 years. Both groups were apparently well educated with the majority ($\approx$90%) having at least some college education which was higher than the general U.S and Indian population. In November 2009 Indian workers on average earned $1.58/h, and U.S workers earned $2.30/h [97]. About 55% of Indian workers reported an annual household income of less than $10000, while 45% of U.S workers reported less than $40000 of annual household income [54, 55]. Meanwhile, 55% of U.S workers live alone or with one more person in the same household and 57% of Indian workers leaving with 3 or more people in the same household.

Crowdworkers from other platforms are rarely studied with the exceptions of Hirth et al. [41], who investigated the locations of workers and requesters of the Microworkers platform, Berg's study [5] who compared MTurk U.S and Indian workers to the CrowdFlower workers, and Peer's study [90] which compared workers from CrowdFlower and Prolific Academic with their colleagues from MTurk.

© Springer International Publishing AG 2018
B. Naderi, *Motivation of Workers on Microtask Crowdsourcing Platforms*, T-Labs
Series in Telecommunication Services, https://doi.org/10.1007/978-3-319-72700-4_3

Although all of these studies are informative, they are based on data collected some years ago which may not be accurate anymore [105] due to the rapid change of the crowdworking population.

In this chapter, the collection and analysis of crowdworkers' demographic data from different platforms is described. It is essentially based on [72], and represents a summary of its relevant parts. The reported studies were designed, conducted, and analyzed by the author except Microworkers (MW) related studies which are reported for the sake of completeness.

3.1 Method

Demographics of crowdworkers from three different crowdsourcing platforms were studied, namely MTurk, Microworkers, and Crowdee. We created a general survey which was later customized for each platform by using their own terminologies (i.e. using "job" in Crowdee instead of "HIT" in MTurk, cf. Sect. 1.1.1). The following items have been considered in the questionnaire:

- **Demographics** (gender, age and size of household). These items were used to characterize the workers in each platform demographically. Results should be considered during platform selection, task design and analysis especially for subjective studies conducted in crowdsourcing [13].
- **Socioeconomic status** (education level, yearly household income, employment status, expenditure purpose of money earned through crowdwork). The second group of items was related to the socioeconomic status of the workers. Previous studies have shown that the primary motivation of workers is monetary, which influences their preferences [73]. The aim was to know how much they rely on their crowdworking income. Workers also specified the expenditure of their crowdworking income (either for primary expenditures such as bills and rent, or secondary expenditures such as hobbies and gadgets).
- **Crowdwork conditions** (weekly time spent on crowdworking, approval rate and number of approved tasks). These items were used to characterize working conditions and attitudes of crowdworkers.

In every survey at least one trapping question [82] was employed to evaluate the reliability of the collected answers later. A trapping question has a straightforward answer which workers should be able to answer correctly without any specific background knowledge. They just need to read the question carefully (cf. Sect. 4.2.1.1). Here workers were asked to indicate how many characters does the word "crowdsourcing" have.

3.2 Data Collection

The demographic studies were conducted separately for each platform in March and April 2016. Table 3.1 summarizes the experimental setup.

For MTurk, two HITs were published in March 2016 with the aim of 100 participants from the U.S and 100 participants from India population of workers (Study 3.1). The HITs were published at 9 AM PDT and within 56 min for U.S workers, and 62 min for Indian workers, all answers were collected. Reliability check questions indicated that ten responses from U.S participants and 29 responses from Indian workers were unreliable. Consequently, the job was extended for Indian workers to gather further data. Altogether, 90 responses from U.S workers and 87 responses from Indian workers were considered reliable and kept for further analysis. The U.S

Table 3.1 Overview of the collected and analysed data. The "Data (acronym)" column reports the origin of the data (survey or MTurk Tracker) as well as the acronym used in tables throughout the rest of the chapter

Platform	Data (acronym)	Continent/ Country	Date range	Valid responses	Rewards ($)	Duration of Study
MTurk	Survey (MTurk U.S)	U.S	Mar. 2016	90	1	56 min
	Survey (MTurk IN)	India	Mar. 2016	87	0.7	62 min
	MTurk Tracker (MTurk U.S 2015-16)	U.S	Apr. 2015 – Jan. 2016	23839	0.05	–
	MTurk Tracker (MTurk IN 2015-16)	India	Apr. 2015 – Jan. 2016	4627	0.05	–
MW	Survey (MW Western)	Europe	Apr. 2016	122	0.8	5 days
		Oceania		12	1.2	48 h
		North America		64	1.2	3 days
	Survey (MW Developing)	South America	Apr. 2016	28	0.48	1 week
		Asia		107	0.46	3 days
		Africa		48	0.48	22 h
Crowdee	Survey (Crowdee)	Western Europe	Mar. 2016	236	€0.8	7 days

and Indian workers were rewarded with $1 and $0.7 respectively. Different rewards were paid as the national minimum wage differs in both countries.[1]

Since Microworkers is globally active and workers are mostly working in their local working time [41], launching a single campaign for all continents obviously causes gathering responses mostly from workers in the same time zone. Thus, different campaigns have been launched for different groups of continents. Moreover, the campaigns were run at minimum speed in the beginning to reach a representative sample of workers in each continent. Rewards for each campaign were set by following the recommendations of the platform. For more details see [72]. Overall data from 474 workers were collected in April 2016 from which 380 were considered reliable. For the analysis two separate groups are created, roughly identified based on the GDP (Gross Domestic Product) of their home countries: (1) a group of developing countries, included in South America, Asia, and Africa, and (2) a group of Western countries, i.e., those included in Europe, Oceania and North America.This grouping was used to compare results from Microworkers to the MTurk, for which U.S (western) and India (developing) workers were analysed separately.

The last demographic study was conducted using the Crowdee platform in March 2016. The survey was split into two jobs, which were published with a aim of 250 participants. In addition to the regular trapping question used in the general survey, participant's birth year was asked in both jobs. Overall 242 workers filled in the survey (i.e. both jobs) completely. All of them answered the trapping question correctly, but six responses were removed as a result of inconsistent answers to the repeated birth year question. As a result, responses of 236 participants were used for further analyses.

In addition to the above-explained studies, a demographic dataset from the MTurk Tracker [21, 55] were included in the analysis. The demographic API of the MTurk Tracker creates a five-item demographic survey job (gender, year of birth, household size, household income, and marital status) in MTurk every 15 min to capture time variability. The survey is compensated with $.05 and workers are also restricted to answer the survey only once a month [53]. Responses are publicly available through the MTurk Tracker API. The dataset contains responses in the time range from April 2015 until January 2016. Among others, 23839 of responses were from U.S crowdworkers and 4627 from Indian crowdworkers.

3.3 Results

In the following the outcomes of data collection on relevant crowd characterization are reported. See [72] for the complete set of results.

---

[1] https://wageindicator.org/main/salary/minimum-wage last accessed 6 February 2017.

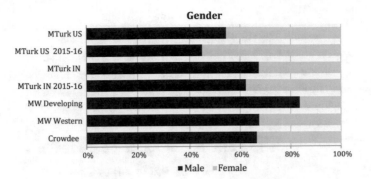

Fig. 3.1 Observed distribution of crowdworkers' *gender* on different platforms, after [72]

3.3.1 Demographics

Gender. Generally, there are less female than male workers (cf. Fig. 3.1). For MTurk, the population of U.S workers is more balanced than Indian workers. Based on the survey data, a larger part of U.S workers are male, which corresponds with other recent surveys [5, 7] but differs from MTurk Tracker data. Meanwhile, the data collected by the survey shows that about 67% of Indian workers are males. This is in line with 71% of India's Internet users' population being male in 2015 [1]. Results from MTurk tracker confirm the relation although there are more females observed than in the survey data.

Similar to the population of Indian workers in MTurk, male workers occupy more than 60% of the population of both Crowdee and Microworkers Western countries workers. The most imbalanced ratio was observed in Microworkers population from developing countries in which there is one female for every five male workers.

Age. The age distributions of participants are illustrated in Fig. 3.2. For both U.S and Indian workers, data collected through MTurk Tracker strongly differs from data obtained through our survey in two age groups namely 18–26 and 41–55 groups in which the survey was answered mostly by the older group. The largest group of U.S workers are 27–32 years old based on the MTurk Tracker data whereas the largest group of participants in the survey were 41–55 years old. Both data sources agreed that the largest group of Indian workers are between 27 and 32 years old.

Microworkers and Crowdee seem to have a younger population of users than MTurk. For Microworkers, the vast majority of workers are 32 or less years old. Workers from Western countries look to be older than their counterparts from developing countries. The young population of Crowdee workers may be due to the fact that the platform is developed and maintained by a university team [81].

Household size. Figure 3.3 illustrates the distribution of household size (including the worker) in the three platforms. For MTurk, most of the U.S workers (>70%) belong to a household with maximum two other persons. A similar trend is observed

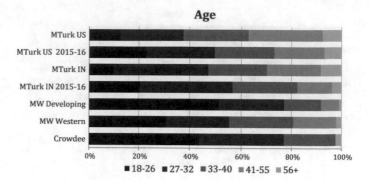

Fig. 3.2 Observed distribution of crowdworkers' *age* on different platforms, after [72]

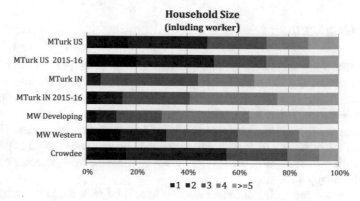

Fig. 3.3 Observed distribution of crowdworkers' *Household size* on different platforms, after [72]

for Western countries in Microworkers (>60). The majority of Crowdee workers (>55%) live either alone or with another person.

Although both data sources do not completely agree about the distribution of household size for Indian MTurk workers, the data clearly indicates that their household sizes are larger than their U.S counterparts. In contrast to workers from western countries, most of the workers from developing countries live together with three persons or more (Indian MTurk workers >56% and Microworkers developing countries workers >69%).

3.3.2   Socioeconomic Status

From the remaining items, the MTurk Tracker demographics contains only the Household income data.

Educational level. Crowdworkers are well educated, as in all platforms, 70% or more of them have at least some college education (cf. Fig. 3.4). Moreover, workers from developing countries achieved significantly higher educated level than their Western counterparts. 57% of the MTurk's Indian workers specified to have a Bachelor's degree and 37% a Master's degree. However, holding a degree is not a good measure of one's foreign language or computer skill in developing countries.[2] Both U.S and Indian population of MTurk workers are more educated than general U.S and Indian population. Workers from Western countries in Microworkers achieved higher education levels than MTurk U.S and Crowdee workers. Crowdee workers mostly have some college education or less (67%). At least some part of them should be still in the middle of their educational path.

Household yearly income. The survey data are similar to the data from MTurk Tracker. About 60% of U.S workers report a household income below $60000 (cf. Fig. 3.5). About 40% of the MTurk Indian population and 60% of Microworkers from developing countries reported a household income below $10000. A Higher level of yearly income was reported by the Microworkers from Western countries and Crowdee workers. Considering the 75% lowest-income group, the Indian MTurk workers yearly earn less than $8000 per capita in their household whereas U.S workers earn less than $20000, and Crowdee worker less than $25000.

Employment status. Independent of their location, a large proportion of the MTurk workers ($> 44\%$) have a full-time job in addition to their crowdsourcing job (cf. Fig. 3.6). Also a proportion of the workers are working part time. A relatively large number of the U.S workers (almost 25%) are keeping house. Most participants from Microworkers, either have a full-time or part-time job besides their crowd work. Compared to the U.S workers of MTurk, Microworkers workers, are more likely to have a part-time job and being students, and less likely to be housekeepers. Crowdee

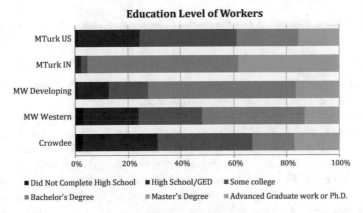

Fig. 3.4 Observed distribution of crowdworkers' *Education level* on different platforms, after [72]

[2]http://www.wsj.com/articles/SB10001424052748703515504576142092863219826 last accessed 16 March 2017.

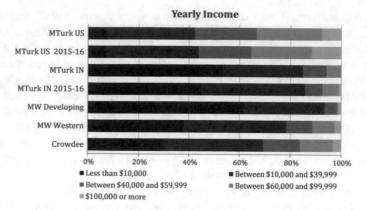

Fig. 3.5 Observed distribution of crowdworkers' *Household income* on different platforms, after [72]

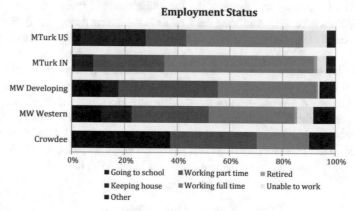

Fig. 3.6 Observed distribution of crowdworkers' *Employment status* on different platforms, after [72]

has the highest population of students which is reflected in the young age of the workers and explains their low level of education.

Expenditure purposes of the money earned through crowdwork. Although similar patterns based on countries of crowdworkers were expected, similarities within platforms are observed. MTurk workers more rely on their crowdwork income for everyday living expenses (like paying bills, gas, groceries etc.) than workers in other platforms (cf. Fig. 3.7). Crowdee and Microworkers workers mostly use their earnings for "secondary" expenses or as pocket change (for hobbies, gadgets, going out etc.). Workers who more rely on their crowdwork income for everyday living expenses may take crowdworking more seriously as it is used to support themselves and even their families.

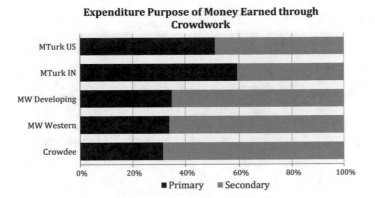

Fig. 3.7 Observed distribution of crowdworkers' *Expenditure purposes of crowdworking income on different platforms*, after [72]

On average, the proportion of workers who use their crowdworking income for primary expenses had the maximum yearly household per capita income of $5400 for Indian, and $15800 for U.S MTurk workers, and $12700 for Crowdee workers. Whereas the group of workers, who use their income for secondary expenses, has on average maximum yearly income per capita of $8800 for Indian, and $24900 for U.S MTurk workers, and $19600 for the Crowdee workers.

3.3.3 Crowdwork Conditions

Time spent on crowdsourcing platforms. Although many of MTurk workers stated to have a full-time or a part-time job, 60% of them additionally spend more than 15 h per week on crowdworking (cf. Fig. 3.8). A large number of them is even working more than 25 h per week (U.S: >37%; India: >47%). A large percentage of workers (≈50%) spend relatively few hours on Microworkers (less than 10) and a smaller but also considerable portion spends more than 25 h (≈25%). For Western workers, the distribution is more skewed towards a smaller number of hours spent on crowdsourcing. In Crowdee workers spend very little time on the platform. This is however due to the smaller number of jobs available on Crowdee compared to MTurk and Microworkers.

Approval rate and number of approved tasks. MTurk workers were asked to report their performance statistics. The *Approval Rate* is a percentage of overall assignments submitted by the worker that have been approved by all requesters. The other statistic indicates the number of assignments submitted by them which were approved by all requesters. Both of these statistics are available in workers profile, but inaccessible by requesters. From 96.7% of U.S and 65% of Indian participants stated to have 98% or higher approval rate. In addition, 21% of Indian workers also

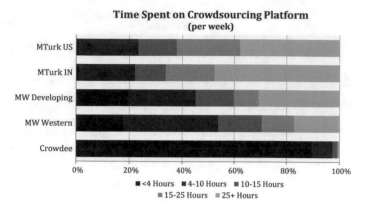

Fig. 3.8 Observed distribution of crowdworkers' *Time spent on crowdsourcing platforms* on different platforms, after [72]

stated to have an approval rate in a range between 95 to 98%. 85% of U.S and Indian workers stated to successfully performed more than 1000 tasks, and even 42% of them stated to have more than 10000 approved assignments.

3.4   Chapter Discussion and Summary

The results show that the demographics, socioeconomic status, and crowdworking conditions of the workers differ considerably between and within platforms. For some items (household size, educational level, and yearly income), the Indian MTurk workers share similarities with the Microworkers from developing countries. Workers from the Western countries shared similarities together, irrespective of the platform they use for crowdworking. Participants from developing countries reported to live in larger households with less yearly income and obtained higher educational levels compared to their Western counterparts. Note that holding a university degree, may not correspond to advanced English language levels in the developing countries as most of the educational programs are taught in their local, or regional languages.

The collected data showed that the time spent for crowdworking relates to expenditure purpose of money, and household per capita income. MTurk workers appear to be more dedicated to crowdwork, use resulting income to support their primary costs, and spending longer hours on the platform, which may suggest higher specialization and possibly efficiency in completing jobs. The population of Crowdee workers and Microworkers from developing countries are younger than the others. Using Crowdee, it is easier to find currently studying participants.

Researchers should consider the observed within and between platform differences of crowdworkers when deciding on the platform to use and selecting their participants. In case that their research results may be influenced by demographics and socioeconomic status of participants, adequate sampling procedures should be considered to avoid potential biases in the results. In addition, the population of crowdworkers is not representative of their country. For instance, in developing countries, basic English language, computer usage knowledge, and the Internet access—which all needed for crowdworking are—associated with higher socioeconomic status.

Despite the fact that crowdsourcing microtask is limited to solve a problem that needs basic human intelligence, results from this study suggest that such a viewpoint potentially can change. First, a large portion of crowdworkers reported having either a part-time or a full-time job. Therefore, they are qualified for certain skills. Second, they are highly educated compared to the population of their countries. Platform providers should consider how to reduce more complicated skillful tasks to smaller microtasks to utilize such an opportunity.

Last but not least, the differences between our survey data and the data obtained from MTurk Tracker indicate that results of one-shot surveys are not necessarily corresponding to the data repeatedly collected in different timestamps over a year. Therefore it is not advisable to generalize demographic results by using a survey job. One solution would be to use two-step jobs, in which the first one should be available for a broad range of participants with a minimum number of questions and posted in different timestamps. Later, a randomly selected sample of participants from the first job should be invited to perform the second job. Meanwhile, as evidence of high-efficiency working habits has been reported, workers may be specialized on a specific type of job. Therefore collected data through survey jobs may not be representative of entire crowdworker communities. Usual survey participants may differ from other workers who normally perform the other type of jobs like annotations, content creation, or information finding.

Chapter 4
How to Measure Motivation?

In this chapter, the development of the Crowdwork Motivation Scale (CWMS) for measuring the general underlying motivation of crowdworkers is described. The CWMS is based on the Self-Determination Theory (SDT) of motivation (cf. Sect. 2.1.1.1) and measures the intrinsic and a spectrum of the extrinsic motivation in the context of crowdworking. For developing the questionnaire, first, a preliminary item pool was created considering a pilot study [85] and validated questionnaires from different domains including work motivation based on the SDT. Later items were customized for the domain of crowdworking by experts. Next, based on the results of an empirical study and confirmatory and exploratory factor analysis some items were dropped which leads to the final questionnaire. Last but not least, the CWMS questionnaire were evaluated in different empirical studies. This chapter is based on [85] and [84]. In the following, each step of developing and validating the CWMS questionnaire is described.

4.1 Initial Item Set

As mentioned in the Chap. 2, SDT suggests that motivation can vary in the degree of being autonomous versus being controlled. Figure 4.1 illustrates the continuum of motivation as proposed by SDT.

To develop an instrument for measuring the spectrum of motivation as suggested by the SDT, we first conducted a pilot study which was focusing on the continuum of the extrinsic motivation. As a result a 12-item questionnaire for measuring extrinsic motivation [85] was developed. Later, we included items for measuring the intrinsic motivation and further refined the items for measuring the extrinsic motivation. The primitive item pool was created based on the results of the pilot study [85], the Intrinsic Motivation Inventory (IMI [75]), the Self-Regulation Questionnaires (SRQ [98]), and the Work Extrinsic Intrinsic Motivation Scale (WEIMS [109]). All of them are

© Springer International Publishing AG 2018
B. Naderi, *Motivation of Workers on Microtask Crowdsourcing Platforms*, T-Labs Series in Telecommunication Services, https://doi.org/10.1007/978-3-319-72700-4_4

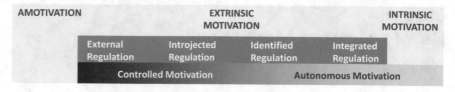

Fig. 4.1 The SDT continuum from amotivation to intrinsic motivation (based on [32])

theoretically grounded in SDT. The IMI focuses on the intrinsic motivation and factors influencing it. It was designed to assess the subjective experience of participants related to a specific activity in the laboratory experiments [75]. It includes the interest and enjoyment subscale which is considered to be the self-reported measure of the intrinsic motivation and several subscales for predicting it (like Perceived Competence, Effort and Importance, and Relatedness). The SRQ focuses on the extrinsic motivation and assesses domain-specific individual differences considering the types of regulation [98]. It differentiates types of regulation based on the degree of internalization. Finally, the WEIMS is an 18-item scale for measuring work motivation based on the SDT. We selected 33 items from the IMI, SRQ and WEIMS scales and adapted them to the domain of crowdworking.

4.2   Study 4.1: Construction of the CWMS Questionnaire

4.2.1   Method

This study was conducted via MTurk for developing the questionnaire focusing on the U.S workers. We decided to use MTurk as it is the must widely used crowdsourcing microtask platform (cf. Sect. 3.1). The questionnaire contained 97 items including the 33 items for assessing the intrinsic and extrinsic motivation, 32 for measuring related constructs (i.e. factors influencing intrinsic motivation like Effort and Perceived Competence), 18 items to collect demographic data, and 14 items to check the reliability of the answers. Participants were asked to indicate the extent to which the items statements are applicable to them on a 7-point Likert scale from 1 "*Does not correspond at all*" or "*Not at all true*" to 7 "*Corresponds exactly*" or "*Very true*". First, the questionnaire was subject to an expert evaluation. The remarked items were revised based on the experts' comments which mainly referred to the wording of the items' statement.

To avoid one particular group of workers to participate in the study and bias the results, we created five jobs with identical content (i.e. the questionnaire) but different required qualification. Based on the taxonomy presented in Sect. 2.1, the type of underlying motivation influences the quality of outcome. Prerequisites were specified based on the *Approval Rate* statistic of the workers' profile to reach a representative sample of participants. The approval rate indicates the percentage of assignments submitted by the worker that have been approved. As a result, five survey jobs were published in MTurk for U.S workers each with a different range of approval rate as a required qualification; namely [0,85], (85,90], (90, 95], (95,98], and (98,100].

Overall, 405 crowdworkers participated in the study. To force participants to answer all questions before submitting their response, the *required* attribute were added to all HTML input tags. However, this feature is not supported by the Safari web browser, which leads to missing data for twelve participants. Removing those cases, 393 completely filled in responses were consolidated.

4.2.1.1 Response Reliability Check

Two measures were employed to check the reliability of the responses. The first measure is based on eight trapping questions [82], which were spread in the questionnaire. Table 4.1 contains the trapping items and shows the range of accepted responses on the 7-point Likert scale. These items act as a gold standard, and they are expected to indicate whether a worker pays adequate attention to the statements. Responses of 42 participants with more than one answer outside the accepted range were rejected.

Furthermore, the *Inconsistency Score* (IS) [82] is used. The IS shows the overall response inconsistency of an individual participant and may be interpreted as an indicator of attentively responding rather than random answering. To use the IS measure, six items from the questionnaire were randomly selected and asked twice in the questionnaire. They are called *consistency check items*. In the post-processing step, for each response the Inconsistency Score was calculated using the normalized

Table 4.1 Trapping questions used in the questionnaire. Participants answered them using 7-point Likert scale: from 1 ("Does not correspond at all" or "Not at all true") to 7 ("Corresponds exactly" or "Very true")

Item	Accepted Response Range in Scale [1,7]
Right now, I am answering a survey in MTurk	[5,7]
If you read this question, please select the answer somewhat true	[4]
I answered all questions in this survey without high concentration; therefore, you should reject my answer	[1,3]
I am fully concentrated and read this text. For this statement, select "Corresponds moderately" to not be rejected	[4]
I think my answer to this task should be rejected, because I did not read this text	[1,3]
I am giving honest answers to this survey	[5,7]
I believe two plus five does not equal nine	[5,7]
Because Mr Xung Illw has personally asked me to work on MTurk	[1,3]

weighted Euclidean distance:

$$IS = \sqrt{\sum_{i=1}^{N} \frac{w_i}{\sum_{j=1}^{N} w_j} (I_i - I_i')^2} \qquad (4.1)$$

where N is the number of consistency check items, I_i is the answer to ith consistency check item by the participant and I_i' is the answer to the duplicate item of I_i by the same participant. w_i is a weight of the i_{th} consistency check item and is calculated similarly to the Inverse Document Frequency [106] known from the Information Retrieval domain, using responses of all participants:

$$w_i = 1 + \log \frac{M}{m_i} \qquad (4.2)$$

where M is the total number of participants and m_i is a number of participants who submitted unequal answers in item i and its repetition, i.e. $|I_i - I_i'| > 0$. When a consistency check item i has a low w_i weight, the item is frequently answered inconsistently by all participants. Thus, given a low w_i, unequal answers in item i and its repetition, does not essentially indicate an inattentive worker but rather a problematic (e.g. an ambiguously worded) item. The distribution of the inconsistency scores of the first 60 responses was used to calculate a cut-off threshold (T) for rejecting or accepting responses:

$$T = Q3(IS) + 1.5 \times IRQ(IS) \qquad (4.3)$$

where $Q3(IS)$ is the 75th percentiles of the inconsistency score distribution and $IQR(IS)$ is the Interquartile Range of that distribution.

Responses of 55 participants were rejected due to the low reliability of their answers detected by one of the approaches mentioned above. In addition, responses of 31 participants, which contained just one wrong answer to the trapping questions, were excluded from further processing. Next, the remaining 307 responses were examined looking for univariate outliers using box plots and standardized scores. As suggested by Tabachnick and Fidell [108], cases with absolute z-score larger than 3.29 were considered to be potential outliers. Responses with more than one potential outlier (detected by either box plot or z-scores) were removed leading to 284 filled in questionnaires remained in the dataset that satisfy our response reliability prerequisite.

The demographic data of participants are reported in Fig. 4.2. Comparing the demographic data from participants in this study with the MTurk Tracker data reported in Sect. 3.3 (MTurk U.S 2015-16), there were no statistically significant difference between gender ($\chi^2(1) = .081$, $p = .776$), age group ($U = 4625$, $z = -.943$, $p = .346$), and yearly income group ($U = 4481$, $z = -1.446$, $p = .148$)

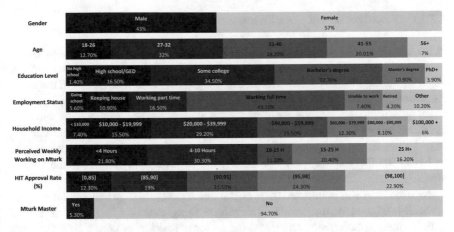

Fig. 4.2 Demographics of participants in Study 4.1 after removing unreliable responses (N = 284)

of both population.[1] In addition, the employed sampling strategy (i.e. publishing five survey jobs, each with different range of approval rate) leads to a fair distribution of approval rate within participants included in the final dataset.[2] Consequently, the dataset is suitable for developing the questionnaire.

4.2.2   Factorial Structure of the Questionnaire

The responses were randomly divided into two groups. The first group contained approximately 40 % of the data (N=114) and was used to construct the questionnaire. The rest was used to validate the final model (cf. Sect. 4.3). A mixed method approach of exploratory and confirmatory factor analysis was used to test the factorial validity of the intended theoretical construct, i.e. whether the underlying motivation of crowdworkers followed the structure of the SDT (cf. Sect. 2.1.1.1). Exploratory factor analyses (EFA) are used when the underlying factor structure is unknown or uncertain; whereas confirmatory factor analysis (CFA) is used when a pre-defined factor structure is known or at least assumed [10]. The development process employed in [114], originally detailed by Homburg and Giering [42], was largely followed.

---

[1]For other fields, data are not available through MTurk Tracker.
[2]Workers with lower approval rate are rear as a high score of this statistic is prerequisite for most of jobs.

4.2.2.1 Construction of the Dimensions

In the first step, each subscale was treated separately. The procedure starts by calcu-
lating the Cronbach's α value, the internal consistency, for each of the six constructs.
Higher values for the Cronbach's α indicate a higher internal consistency of the scale.
Accordingly, items that lowered the Cronbach's α value of their corresponding construct were removed. Results are presented in Table 4.2. The Cronbach's α value for
EXT was below 0.7 which considered as the minimum by Homburg and Giering
[42]. Values substantially lower indicate an unreliable scale. The large values of α
on ENJO and INTRO may partly be due to their large number of items as the α value
also depends on the number of items [28].

Next, an EFA was conducted for each dimension to evaluate whether all items theoretically assigned to a dimension actually form a single factor. Maximum Likelihood
with Varimax rotation was employed for factor extraction to create an orthogonal
factor structure. Each resulting factor should at least explain 50% of the variance.
Consequently, four items were removed and 21 items remained in the item pool
(Table 4.3). Note that a positive degree of freedom is required to calculate the χ^2-
goodness of fit statistic which was achieved whenever the dimension contains four
or more items. The non-significant χ^2-goodness of fit indicates that the one factorial
structure is a good description of data.

Table 4.2 Cronbach's α value of all constructs. Removing some items increased the α values

Scale	Cronbach's α Before/After	N of Items Before/After
Interest and enjoyment (ENJO)	.922 / .922	7 / 7
Identified regulation (IDEN)	.765 / .815	5 / 3
Integrated regulation (INTEG)	.775 / .775	2 / 2
Introjected regulation (INTRO)	.893 /.894	8 / 7
External regulation (EXT)	.181 /.657	8 / 4
Amotivation (AMO)	.665 /.782	3 / 2

Table 4.3 Exploratory factor analyses for all dimensions using Maximum Likelihood with Varimax rotation

Scale	N of Items	Explained variance	χ^2	df	p
ENJO	5	69.304	9.71	5	.08
IDEN	3	72.959	n.a	n.a	n.a
INTEG	2	81.705	n.a	n.a	n.a
INTRO	6	64.358	9.35	9	.41
EXT	3	67.104	n.a	n.a	n.a
AMO	2	82.21	n.a	n.a	n.a

The procedure was followed by conducting confirmatory factor analyses for each construct using IBM® SPSS® AMOS (version 22). In CFA, models are specified based on previous theoretical assumptions. Then it is tested whether the hypothesized factor structure of the model fits the sample data [10]. A single factor structure (i.e. measurement model) for each construct was created and evaluated. The evaluation of the model-fit should be based on several criteria which particularly focus of adequacy of (1) the parameter estimate, and (2) the model as a whole [10] . The following criteria, suggested by [42, 52, 108] and used by [115], were employed:

Item reliability (IR) $\geq$ 0.4 IR is the estimated R^2 (squared standardized factor loading) and explains how much variance of each item is accounted for by the factor the item is loading on [115]. It has a range of [0, 1] and higher values indicate higher reliability [114].

The rest of criteria applies to the model as a whole:

Composite reliability (CR) $\geq$ 0.6 CR is a measure of the overall reliability of a factor which shows how well it can be measured via the items loading on it [115]. It has a range of [0, 1] in which higher values indicate higher reliability.

Adjusted χ^2 by the degrees of freedom (df) $(\frac{\chi^2}{df}) \leq 3$. The χ^2-test examines the null hypothesis whether "the specification of the factor loadings, factor variances and covariances, and error variance for the model under study are valid" [10]. A nonsignificant χ^2-test indicates a good fit, however, the p—value depends to the sample size [52, 108]. As p-values are decreasing with increasing sample sizes, it has been advised that the fit of the model can more plausibly be estimated by using the χ^2 statistics adjusted by its degrees of freedom [52].

Comparative fit index (CFI) $\geq$ 0.95. It compares the hypothesized model with an independence model (the model in which all variables are unrelated) using a non-central χ^2-distribution [52, 108]. When the proposed model is better than the independence model, the CFI increases, i.e. the fit gets better [114]. It has a range of [0, 1]; values of 0.95 and higher show an excellent fit, i.e. lower Type II error rates with acceptable costs of Type I error rates [50].

Average variance extracted (AVE) $\geq$ 0.5. The AVE indicates the relation between the variance explained by the factor and the variance resulting from the measurement error [114]. It has a range of [0, 1] and higher values indicate better fit. The values larger than 0.5 show that the variance explained by the factor is greater than the variance due to the measurement error.

Root mean square error of approximation (RMSEA) $\leq$ 0.08. The RMSEA or "badness-of-fit" index has a range of [0, 1]. It considers the error of approximation in the population and becomes zero when the predictions of the model match the data perfectly [10, 52]. A value smaller than 0.08 indicates a good fit and a value lower than 0.05 indicates an excellent fit [10].

All dimensions were tested considering the above-mentioned criteria (Table 4.4). In case that the model is saturated (i.e. the model has zero degrees of freedom) some of the explained criteria cannot be calculated. It was the case for four constructs. Consequently, only IR and CR were calculated for those dimensions. For each, INTRO and EXT, one item was excluded due to a low IR score. As a result, overall 19 items remained in the item pool.

Table 4.4 Fit indices for all dimensions of the intermediate item set

Scale	N of Items	IR Min./Max	CR	$\frac{\chi^2}{df}$ (p)	CFI	RSMEA
ENJO	5	.49 /.82	.89	1.99 (.08)	.99	.088
IDEN	3	.57 /.62	.81	n.a	n.a	n.a
INTEG	2	.59 /.69	.78	n.a	n.a	n.a
INTRO	5	.44 /.69	.88	1.63 (.15)	.99	.07
EXT	2	.59 /.89	.85	n.a	n.a	n.a
AMO	2	.59 /.7	.79	n.a	n.a	n.a

4.2.2.2 Construction of the Global Model

The structure of the global model was examined to investigate if the proposed factor structure matches the data. In addition to the above-explained criteria, two further measures were considered in this step: convergent validity and discriminant validity. The convergent validity is the degree to which a construct converges with other theoretically similar constructs [69]. It is obtained when the IR criteria and the AVE criteria are satisfied. The discriminant validity is the degree to which a construct differs from other theoretically dissimilar constructs [69]. In EFA, it is obtained when constructs are not correlating more than 0.7, and in CFA the square root of the AVE should be larger than the inter-factor correlations. EFA and CFA for the questionnaire as a whole were conducted.

In EFA, the Maximum-Likelihood factor analysis with the Promax rotation were calculated. A promax rotation was performed rather than Varimax as Promax allows the factors to correlate [115]. Based on the SDT, subscales (i.e. factors) presenting different levels of internalization should positively correlate (cf. Fig. 4.1). Within the EFA, at first, a six-factor solution was examined as based on the theory six dimensions were expected. Results show that INTEG items and IDEN items were loading on the same factor. Based on the SDT, the identified and the integrated regulations are both highly internalized types of extrinsic motivation, and together with intrinsic motivation they form autonomous motivation (cf. Sect. 2.1.1.1). Therefore, they may not be recognized as independent types of motivational regulation in the crowd-working domain. This assumption was tested by extracting five factors. In addition, two more items, one from INTRO and one from ENJO, were removed due to low loadings (< 0.4) on the assumed factor and high cross-loading (> 0.3) on other factors. The χ^2-goodness-of-fit indicates that the hypothesized five-factor structure does not differ significantly from the data structure, $\chi^2(61, N = 114) = 59.196$, $p = .542$. As a result, the INTEG dimension was merged with the IDEN dimension. The pattern matrix of the five-factor solution is reported in Table 4.5. Note that EXT9 and AMO17 are Heywood cases (factor loading greater than 1.0) which may have occurred due to the relatively small size of the dataset [114].

Table 4.5 Factor loadings for intermediate item set based on a Maximum Likelihood Factor Analysis with Promax rotation ($KMO = .821$, Bartlett's test: $\chi^2(136, N = 114) = 947.431$, $p < .001$)

Item	Factor 1 ENJO	Factor 2 IDEN	Factor 3 INTRO	Factor 4 EXT	Factor 5 AMO
ENJO1	.925				
ENJO3_R	.697				
ENJO5	.877				
ENJO8	.728				
IDEN1		.822			
IDEN7		.710			
IDEN14		.762			
INTEG18		.759			
INTEG5		.487			
INTRO41			.725		
INTRO43			.682		
INTRO45			.547		
INTRO46			.975		
EXT2				.646	
EXT9				1.012	
AMO12					.568
AMO17					1.009

Note. Factor loadings $< .3$ are omitted

Furthermore, results of the EFA show sufficient convergent and discriminant validity: factors loadings were all above the recommended minimum threshold; the highest correlation value in the correlation matrix was 0.568, which is well below 0.7. Also, there were no cross-loadings above 0.3. The five-factor model explained a total variance of 64%.

Next, the complete questionnaire was modeled using a CFA in IBM® AMOS. The criteria explained above were checked to see how the proposed structure meets the data. After removing the item INTEG5, the model (Fig. 4.3) shows a good fit: $\chi^2(95, N = 114) = 110.12$, $p = .138$, $\frac{\chi^2}{df} = 1.159$, $RMSEA = .038$, $CFI = .981$, $IR_{min} = .46$, $IR_{max} = .91$.

Also, for each factor, the composite reliability was above the minimum threshold of 0.7 and the AVE was above 0.5, indicating a satisfactory convergent validity (Table 4.6). All factors showed adequate discriminant validity (i.e. the inter-factor correlations are lower than the square root of the AVE). Accordingly, this 16-item solution, which is measuring underlying workers motivation in crowdsourcing platform on five dimensions, was accepted for the final questionnaire.

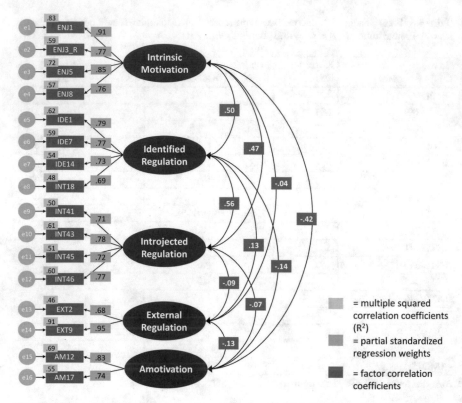

Fig. 4.3 First-order CFA model for the Crowdsourcing Work Motivation Scale (CWMS) [84]

Table 4.6 Composite Reliability (CR), Average Variance Extracted (AVE) and inter-factor correlations using the training dataset

	CR	AVE	Inter-factor correlations and square root of AVEs				
			ENJO	IDEN	INTRO	EXT	AMO
ENJO	.892	.767	**.822**				
IDEN	.833	.556	.496	**.746**			
INTRO	.833	.555	.468	.559	**.745**		
EXT	.809	.685	-.036	.126	-.088	**.828**	
AMO	.764	.618	-.415	-.142	-.074	-.133	**.786**

Square root of AVEs are on the diagonal

Table 4.7 Fit statistics for the new datasets

	Study 4.1 Test dataset U.S workers	Study 3.1 U.S workers	Study 3.1 Indian workers
N	170	90	86
$\chi^2 (df = 95)$	243.731, $(p < .001)$	141.491, $(p = .001)$	136.575, $(p = .003)$
$\frac{\chi^2}{df}$	2.566	1.489	1.438
RMSEA	.096	.074	.072
CFI	.909	.919	.944
IR_{min}	.36	.3	$.18^a$
IR_{max}	.99	.85	.9
CR_{min}	.71	.65	.8

[a]Except ENJO3, all the other IR values were above .45

4.3 Validation of the Questionnaire

Three datasets have been used for validating the questionnaire and ensuring that the model structure is not sample-dependent. The test dataset[3] (cf. Sect. 4.2.1) with 170 participants from the Study 3.1 and two other datasets from the Study 3.1, in which CWMS questionnaire were reliably answered by 90 U.S and 86 Indian respondents. In the Study 3.1, no other restriction than location was applied. The model fit indices for all three datasets are presented in Table 4.7.

Although the null hypothesizes are rejected, the model's fit statistics are in accordance or close to the suggested criteria. Tables 4.8, 4.9 and 4.10 present the descriptive statistics for the CWMS using the full dataset of the Study 4.1 and the data of U.S and Indian participants of the Study 3.1. All subscales of CWMS have high reliabilities (Cronbach's $\alpha > 0.7$) in all three datasets with an exception of External motivation for U.S population of the Study 3.1 (Cronbach's $\alpha = .619$).

Finally, Table 4.11 contains the list of the items remaining in the final CWMS questionnaire.

4.4 Chapter Discussion and Summary

This chapter described the development and validation of the Crowdsourcing Work Motivation Scale (CWMS), for measuring the general underlying intrinsic and a spectrum of extrinsic motivation in the context of crowdworking. The developed instrument is based on the assumptions of the SDT [32, 100] and the results indicate that the CWMS provides reliable assessments of the dimensions proposed in SDT.

---

[3]The collected data in the Study 4.1 were splited to training and test dataset.

Table 4.8 Descriptive statistics for the CWMS, Cronbach's α with 95 % confidence intervals, using training and test dataset of Study 4.1

	Mean	SD	Correlations and Cronbach's α					Cronbach's α	
			Intrinsic	Identified	Introjected	External	Amotivation	95%	CI
Intrinsic	5.02	1.39	.891					.87	.91
Identified	3.64	1.67	.473[b]	.824				.79	.86
Introjected	3.46	1.8	.463[b]	.494[b]	.873			.85	.9
External	6.68	.58	−.082	.072	−.126[a]	.791		.74	.83
Amotivation	1.74	1.08	−.265[b]	−.005	.096	−.158[b]	.729	.660	.790

[a]Correlation is significant at the .05 level (2-tailed)
[b]Correlation is significant at the .01 level (2-tailed)
Cronbach's α are on the diagonal

Table 4.9 Descriptive statistics for the CWMS, Cronbach's α with 95% confidence intervals, using training and test dataset of Study 3.1-U.S workers

	Mean	SD	Correlations and Cronbach's α					Cronbach's α	
			Intrinsic	Identified	Introjected	External	Amotivation	95%	CI
Intrinsic	3.37	1.46	.889					.85	.92
Identified	4.22	1.34	.224[a]	.743				.64	.82
Introjected	2.01	1.14	.353[b]	.296[b]	.812			.74	.87
External	6.55	.62	−.295	−.016	−.361[b]	.619		.42	.75
Amotivation	1.68	1.14	−.157	.068	.242[a]	−.328[b]	.795	.69	.87

[a] Correlation is significant at the .05 level (2-tailed)
[b] Correlation is significant at the .01 level (2-tailed)
Cronbach's α are on the diagonal

Table 4.10 Descriptive statistics for the CWMS, Cronbach's α with 95% confidence intervals, using training and test dataset of Study 3.1-Indian workers

| | Mean | SD | Correlations and Cronbach's α | | | | | Cronbach's α | |
			Intrinsic	Identified	Introjected	External	Amotivation	95%	CI
Intrinsic	5.67	1.26	.85					.79	.9
Identified	5.05	1.38	.617[b]	.857				.8	.9
Introjected	4.55	1.7	.491[b]	.668[b]	.861			.81	.9
External	6.26	.9	.084	.164	.058	.788		.68	.86
Amotivation	2.61	1.64	−.176	.143	.252[a]	−.137	.775	.66	.85

[a]Correlation is significant at the .05 level (2-tailed)
[b]Correlation is significant at the .01 level (2-tailed)
Cronbach's α are on the diagonal

Table 4.11 List of final items for CWMS

For each of the following statements, please indicate how true it is for you and to what extent each corresponds to the reasons why you are presently performing tasks in MTurk

Intrinsic motivation	
ENJO1	I am enjoying doing tasks in MTurk very much
ENJO3^R	I think tasks I do in MTurk are boring
ENJO5	I would describe HITs I do in MTurk as very interesting
ENJO8	I perform HITs in MTurk, because I derive much pleasure from performing task
Identified regulation	
IDEN1	Because this is the type of work I chose to do to attain a certain lifestyle
IDEN7	Because I chose this type of work to attain my career goals
IDEN14	Because it is the type of work I have chosen to attain certain important objectives
INTEG18	Because this job is a part of my life
Introjected regulation	
INTRO41	Because I want the requesters and others to think I'm smart and a good worker
INTRO43	Because I want people (requesters or others) to like me
INTRO45	I am strongly motivated by the recognition I can earn from other people
INTRO46	I want other people to find out how good I really can be at my work
External regulation	
EXT2	For the income it provides me
EXT9	Because it allows me to earn money
Amotivation	
AMO12	I don't know why, we are provided with unrealistic working conditions
AMO17	I don't know, too much is expected of us

^{R}Reverse-score Item.
Items should be used in combination with a 7-point Likert scale ranging from 1 ('Does not correspond at all') to 7 ('Corresponds exactly'). After reversing the marked item, the value of each dimension should be calculated by averaging the score of items loading on that dimension

Amotivation shows negative or low correlations with the other dimensions. This is in line with the SDT; here, amotivation is defined as a situation when both, extrinsic and intrinsic motivation, are being absent. In addition, the negative and low correlations between external regulation and internalized motivation are consistent with the SDT. Our results suggest that in context of crowdworking the dimensions integrated and introjected regulation should be merged. Similar conclusions have been reported in [85] and in other domains as well (e.g. [109]). Ultimately, SDT assumes that these two are both part of the autonomous motivation. Moreover, external regulation has a high mean value and a very low standard deviation which indicates a rather low variance for the items loading on this dimension. It could be due to either the items' wording [33], or the external motivation being extremely important in the paid crowdsourcing domain, which is not unexpected. In further research, those items should be closely investigated.

Furthermore, it should be considered that the instrument measures the underlying motivation of general crowdworking. Therefore, future research on user characteristics and platform-dependent factors can utilize this tool to investigate the relative level of influence of different user and platform dependent factors on worker's motivation. Platform providers can measure the effect of any modification in their platform, including the incentives frameworks, on the type and amount of workers motivation. As CET suggests, the environment can facilitate or forestall intrinsic motivation and internalization of extrinsic motivation by supporting versus thwarting the basic psychological needs proposed in the SDT (i.e. autonomy, competence and relatedness, [99]). In addition, the underlying motivation of crowdworking may influence the tasks selection strategies of crowdworkers i.e. task-specific motivation. It might be a case that workers with higher underlying controlled motivation, look for tasks with higher payout and those who are more autonomously motivated select the interesting tasks more often. The next chapter investigates the task selection strategies by crowdworkers and studies which attributes, including underlying crowdworking motivation, influenc them.

Chapter 5
What Determines Task Selection Strategies?

In this chapter, five studies are described aiming to answer questions about factors influencing the task choice of crowdworkers. Accepting to perform a microtask indicates that corresponding motives exist and are more powerful than obstacles which lead to enough motivation for performing the task. The first study aimed to find out the relevant factors (motives and obstacles) that are influencing the workers' task choice. The next study intended to find adequate instruments for measuring these factors. Selected instruments were used in the third study to introduce a model for predicting workers' *acceptance* decision. The model focused on the task-related attributes. In study four, the relative importance of all factors (including non task-related) were evaluated, using different worker groups. The last study provides a model for predicting the expected workload (one of task-related attributes) from task design, and applied the predicted value in the acceptance model.

5.1 Study 5.1: What are the Influencing Factors?

5.1.1 Method

A qualitative study, using MTurk, was conducted in which invited workers are asked to assess another HIT. They specified whether they would perform the presented HIT (assuming it is available for them) and explained *why* in an open-end question. Besides this, the workers answered further questions about the given HIT including rating the difficulty level, enjoyableness, promptness of reward (on a five-point Likert scale), and if they know the requester (binary answer). Sixty active HITs from MTurk were randomly selected and included in the test dataset. They include different HIT types with rewards range from \$0.01 to \$1.5 ($M = 0.17$, $SD = 0.28$). HITs were

© Springer International Publishing AG 2018
B. Naderi, *Motivation of Workers on Microtask Crowdsourcing Platforms*, T-Labs Series in Telecommunication Services, https://doi.org/10.1007/978-3-319-72700-4_5

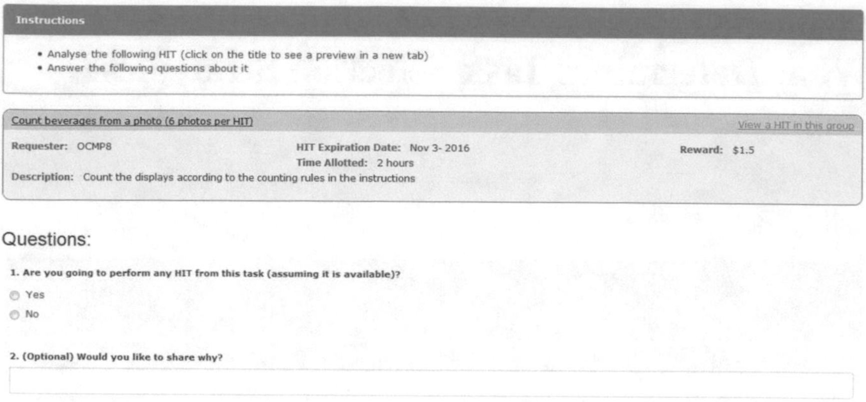

Fig. 5.1 Microtask used in the Study 5.1

presented by their metadata including title, requester, description, expiration date, the allotted time, reward and a link to the real HIT in MTurk (cf. Fig. 5.1).

Each HIT was assessed by five different workers. Workers from the U.S, who hold "Master" qualification of MTurk, were eligible to participate in the study. The Master qualification is granted to workers by MTurk who "consistently completing HITs of a certain type with a high degree of accuracy across a variety of Requesters". This restriction was applied since we aim to understand what factors influence the preference of professional crowdworkers. Finally, workers were compensated with $0.05 for each assignment and assured an extra bonus of $0.05 per assignment when they carried out all available tasks (i.e. evaluating all sixty HITs in the dataset). The study completed within four hours with fourteen participants who assessed all HITs in the dataset.

5.1.2   Results

In 43% of responses, workers gave the reason of their decisions in free text format. Later in the laboratory, three different coders labeled the free text inputs. Overall 34 different labels from three categories were used by coders reaching Fleiss' Kappa of 0.62 which indicates substantial inter-rater reliability [66]. The most used labels are presented in Table 5.1.

Just in 9.9% of times workers specified their willingness to take the given task. Therefore, the dataset is not suitable for training a classifier which can predict workers' decision.

Table 5.1 Labels used for explaining "why" workers do (not) take a micro-task

Class	Sample label	Count	Percentage (%)
Pay	Low/good, reasonable, worth, quick pay	305	47
Requester	Known, ratings	141	22
Task		206	32
... difficulty	Difficult, easy, unequal difficulty	87	
... complexity	Too involving, complicated, confusing, learn much	39	
... interestingness	Uninteresting, tedious	20	
... bad design	Slow load, huge instruction	27	
... type	Type	33	

5.1.3   Discussion

Workers mentioned the payment, requester, and task attributes when explaining their decision on (not) performing the given HIT.

Low Payment was the most used label (238 times), and probably one of the reasons why mostly decided against performing the given HIT. Considering the requesters, workers mostly mentioned to the Turkopticon ratings (cf. Sect. 1.1.1.1). Workers more referred to their negative personal experiences (78%) with requesters than positive ones.

Task-related labels were used in 32% of times. To our understanding, difficulty refers to the size of the task, an amount of activity and required effort. Complexity refers to including many steps and details. Interestingness was used in the negative form either due to interestingness being a *hygiene factor*, referring to the Herzberg's Two-factor theory of motivation (cf. Sect. 2.2), or simply caused by a too small dataset not including interesting HITs. Type of the HIT shows that professional crowdworkers may specialize in particular types of microtask.

A point of criticism about the study is that no preview of the presented HIT was given. In 30% of assignments, workers used the provided HTML link to visit the real HIT's page. However, it could be a case that the presented HIT was finished by other workers and at that point of time it was not available anymore. Therefore, the given judgments were mostly based on the meta-data provided in the assignment.

Another criticism applies to the dataset, as in 90.01% of cases workers decided against performing the presented HIT. Therefore the dataset was not representative. Meanwhile, in the rest of cases, workers did not agree together as well. Although it shows that workers have different expectations and preferences, it should also be considered that no reliability check method was used in this study.

To overcome the mentioned limitations and verify the qualitative results from Study 5.1, the following studies were conducted.

5.2   Study 5.2: How to Measure the Task-Related Factors?

The task-related attributes mentioned by crowdworkers in Study 5.1 like difficulty, complexity and bad design influence the *workload* associated to the microtask. This pilot study aimed to integrate a workload measurement technique and test the study design at a small scale. Hart [36] defines workload as follows:

> Workload represents the cost of accomplishing mission requirements for the human operator.

Real and perceived workload differ while the latter influences the expected compensation. Humans who are overloaded with tasks, work in haste, show poor accuracy and become frustrated [11]. Meanwhile, low workload has also been related to high error rates, frustration, and fatigue [11].

A fair balance between workload and compensation leads to increasing productivity and throughput rate [21]. Beside other factors workload influences the job choice decisions.

Different methods are used for measuring workload including performance measures (e.g. measuring speed and accuracy), subjective measurements, and physiological measurements (e.g. heart rate, respiratory rate, eye blink rate, and brain activity) [11, 76]. A comprehensive review is given by Miller in [76]. For subjective assessment of workload, participants first perform the target task and afterwards fill in a questionnaire containing measurement items which reflect their perceived workload. We aim to measure workers' estimation of workload for the given task, rather than perceived workload. Therefore, neither performance nor physiological measures are applicable. But the subjective measurements can be used without performing the task, to measure *estimated workload*. The subjective measurements can be divided into unidimensional and multidimensional measures depending on the scale used for rating. Unidimensional scales are simple and fast and in some cases more sensitive than the multidimensional scale [18]. The multidimensional workload assessment scales are more complex and are used for diagnostic [18], e.g. SWAT [92] consisting of three dimensions and NASA TLX with six dimensions [36, 37].

In this study the RSME-scale (*Rating Scale Mental Effort* [121]) and the NASA TLX (*NASA Task Load Index* [37]) were employed.

RSME. The RSME (also known as SMEQ) developed by Zijlstra [121]· is a lightweight unidimensional instrument shown to have excellent psychometrics properties [18, 40, 101]. Participants rate invested effort of the task on a continuous vertical line ranging from 0 to 150 with indications in every 10 points. Nine anchor points are also given starting from slightly above 0, with the statement "Absolutely no effort", to slightly above 110 labeled by "Extreme effort". It leaves a substantial rating space above "Extreme effort" which may be used by participants (cf. Fig. 5.2). Contradictory to its name, the RSME-scale is often used to measure the overall effort rather than the mental workload [18, 40, 101]. We employed the RSME by asking participants to indicate their *estimation of overall effort* for completing a given HIT.

NASA TLX. The scale has been developed in 1986 by Hart [37] and is one of the most widely used multidimensional scales for measuring the workload [36].

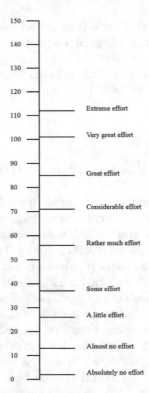

Fig. 5.2 RSME, rating scale mental effort; after Zijlstra, [120]

It consists of six subscales: mental demand, physical demand, temporal demand, effort, performance, and frustration. Ratings for each subscale should be obtained using a twenty-step bipolar scale (cf. Fig. 5.3). It is assumed that some combination of them are likely to represent the overall workload experienced by the participant [36]. The TLX scale uses an individualized weighting scheme for each participant since individuals perceive the influence of each subscale on their overall experience of workload differently [37]. The weights should be calculated using a set of simple pair comparison between six subscales. In each comparison, a participant should decide which dimension is more contributing to their personal definition of workload for the given task [36]. Thus, the TLX result is tailored to an individual definition of workload.

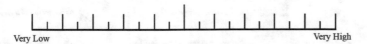

Fig. 5.3 Twenty-step bipolar scale used for rating six dimensions of TLX

A simplified version of TLX scale is called NASA-RTLX (NASA Raw Task Load Index) scale in which the individual weightings are discarded. Ratings of the six subscales are averaged to calculate the RTLX score. Different studies reported a correlation above $r = 0.95$ between TLX and RTLX scores [18, 76].

5.2.1   Method

This study considered to be a pilot study and consisted of three jobs created in MTurk (cf. Fig. 5.4).

Screening job. It contains four demographic questions rewarded by $0.01 and was available for U.S workers who had 100 or more approved HITs. It was published in two different days each time asking for 200 participants. Overall it took ten hours to collect 400 responses. Participants spent on average 1 min to fill in the survey. All participants answered a simple trapping question included in the job correctly. 150 individuals were randomly selected and invited to the next step.

Introductory job. It consists of two parts. In the first part, workers were introduced to TLX subscales by reading the definitions following by a simple single-choice quiz regarding the definition of temporal demand. The second part contains fifteen pairwise comparison questions referring to the six subscales of TLX. In each question, participants should indicate which of two subscales is more contributing to their personal experience of the workload when performing a typical HIT in MTurk. The job was compensated with $0.40. From the invited group of workers, 106 individuals took part in the introductory job. They all answered the simple quiz correctly and were qualified to perform the next job.

HIT rating job. In this job qualified workers assessed a given HIT based on its screenshot and meta-data (i.e. title, description, the allotted time, rewards, HITs available). Figure 5.5 illustrates how an example HIT was represented in this job. Thirty-one HITs were collected from MTurk with rewards range from $0 to $1 ($M = 0.17, SD = 0.29$) excluding one HIT with exceptionally high payout of $8.59. Each of them was assessed by ten individuals. Participants estimated the workload associated with each HIT by rating the six subscales of TLX and one item of RSME. Furthermore, workers assessed other aspects of each HIT namely interestingness, the amount of fair reward, and how likely is it for them to work on the given HIT (hereafter referred to as "Acceptance"). Participants rated each item on twenty-step

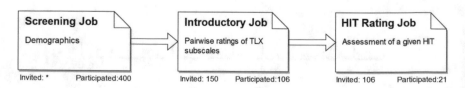

Fig. 5.4 Procedure of Study 5.2

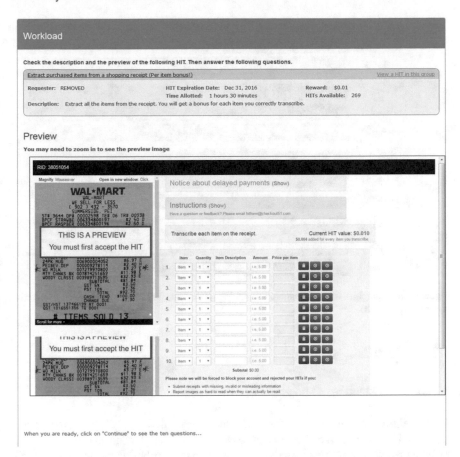

Fig. 5.5 Presentation of a HIT in the HIT rating job as seen by the worker

bipolar scales. To encourage workers to rate more HITs, a bonus payment of $0.05 was added to the base reward of $0.10 per assignment if they performed more than 20 assignments. The maximum of allowed assignments per worker was 30. Twenty-one workers assessed between 1 to 30 HITs ($M = 13$, $SD = 12$). No trapping questions were employed in this study. However 59 observations were discarded as one or more scales were not rated. Further investigations revealed that previews of two HITs were not shown properly due to technical problems. Demographics of participants are reported in Fig. 5.6.

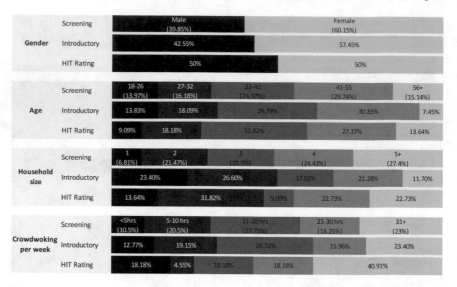

Fig. 5.6 Demographics of participants in screening (N = 400), introductory (N = 106), and HIT rating (N = 21) jobs

5.2.2   Results

5.2.2.1   Workload

TLX weights. Mean weights for TLX subscales rated by 106 participants are reported in Fig. 5.7. Weights distributions were tested for normality using the Kolmogorov-Smirnov Test. Analysis showed that distribution of weights for all subscales significantly deviate from the normal distribution. There was a statistically significant difference between weights assigned to TLX subscales ($\chi^2(5) = 132.361$, $p < .001$). A posthoc signed-rank test showed that the Physical Demand and the Frustration dimensions are weighted significantly lower than the other subscales. No correlation was shown between weightings of subscales and workers characteristics.

  Workload. As expected, TLX and RTLX highly correlated for each observation (Pearson's $r(251) = .959$, $p < 0.001$). The overall effort (i.e. ratings from RSME) was also significantly correlating with both (Pearson's $r_{TLX}(251) = .698$, $p < 0.001$, $r_{RTLX}(251) = .627$, $p < 0.001$).[1] The mean workload score for each HIT was calculated as the arithmetic mean over collected observations for that HIT. The mean workload score calculated based on TLX values ranges from 8.14 to 62.13 ($M = 32.75$, $SD = 12.4$) with standard deviation from 7.69 to 30.44 ($M = 15.38$, $SD = 5.06$). Calculations based on RTLX leads to a range of mean workload score

---

[1] The RSME ratings also significantly correlate with TLX subscales: $r_{effort}(251) = .832$, $p < 0.001$, $r_{MentalD.}(251) = .52$, $p < 0.001$, $r_{PhysicalD.}(251) = .5$, $p < 0.001$, $r_{Frustration}(251) = .412$, $p < 0.001$.

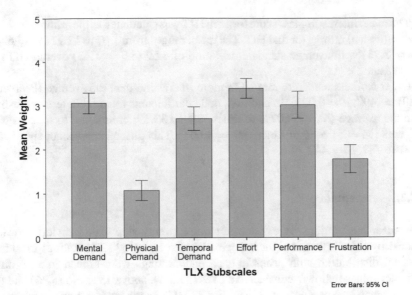

Fig. 5.7 Contribution of TLX dimensions on percieved workload of a typical HIT in MTurk

from 7.98 to 59 ($M = 31.88, SD = 12.26$) with standard deviation between 8.12 and 23.83 ($M = 15.33, SD = 4.14$). The mean overall effort score ranged from 25.22 to 85.98 ($M = 51.82, SD = 15$) with standard deviation between 10.69 to 43.17 ($M = 27.67, SD = 7.7$).

Mean workload score calculated based on the TLX and RTLX were mostly identical (Pearson's $r(29) = .99$, $p < 0.001$) and both significantly correlated with the mean overall effort score (Pearson's $r_{TLX}(29) = .92$, $p < 0.001$, Pearson's $r_{RTLX}(29) = .91$, $p < 0.001$). Older participants tend to report less overall effort for the given HITs (Pearson's $r(20) = -.42$, $p = 0.08$). No other correlation was shown between different estimations of workload and workers characteristics. The overall effort also significantly correlates with allocated reward (Pearson's $r(29) = .48$, $p = 0.008$).

5.2.2.2 Interestingness

The interestingness ratings ranged from 0 to 20 ($M = 6.59, SD = 5.6$). There was a small correlation between the individual interestingness ratings and task's payout (Pearson's $r(251) = .123, p = 0.052$). No other correlation with either tasks properties[2] or worker's characteristics was discovered.

---

[2]A small significant correlation to level of frustration was observed: Pearson's $r(251) = -.191$, $p = 0.002$.

The mean interestingness score for each HIT was calculated as the arithmetic mean over individual ratings for that HIT. The mean ranges from 1.67 to 12.22 ($M = 6.49$, $SD = 3.21$) with average standard deviation of 4.69 ($SD = 1.43$) overall HITs in the dataset.

Older workers tended to rate HITs more interesting than the average (Pearson's $r(20) = .492, p = 0.028$). In addition, male participants rated HITs less interesting than the average (M = 63.07% of times, $SD = 33.19$) comparing to females participants (M = 51.36% of times, $SD = 33.04$). This difference was not significant $t(18) = 787, p = .442$.

5.2.2.3 Acceptance

The Acceptance ratings were collected using a twenty-step scale with left indicator labeled by "Yes" (0) and right indicator labeled by "No" (20). Ratings created a bimodal distribution with peaks in each end. Ratings were binned in an ordinal variable: values bellow 7 considered as Yes (31.5%), above 13 as No (57%) and the middle range as Not Known (11.6%). For 22 HITs in the dataset, participants often (>50% of times) decided against performing them. In nine HITs more than 80% agreed to do not perform that HIT whereas just in three HITs more than 80% of workers agreed to take them.

A significant negative correlation between HIT's mean workload score and the acceptance rate were observed (Pearson's $r(29) = -.404, p = 0.03$). Considering user characteristics, none of participants' age, household size, and level of working hours related to their acceptance rating. Female participants tended to accept more HIT ($M = 37.33, SD = 20$) than male workers ($M = 27.1, SD = 8.28$) on average. However, the difference was not significant $t(11) = 1.072, p = .307$.[3]

Prediction. A logistic regression was performed to determine the effect of task-based (i.e. rewards, interestingness, workload) and user characteristics (working hours, age, gender and household size) on likelihood that participants decide to accept the HIT. For interestingness and workload both individual ratings and over-all means were considered. In addition, estimated *profitability* as a ratio of rewards to individual ratings of workload was calculated and considered as predictor. An intercept-only model (baseline) predicts the decision of workers with 64.4% accuracy. In contrast, the calculated model (Table 5.2) correctly classified 83.3% of cases (sensitivity: 74.7%, specificity: 88.1%) and explained 59.8% (Nagelkerke R^2) of the variance in the workers' decisions. The Hosmer and Lemeshow test[4] suggest that the model was fit to the data well $\chi^2(8) = 8.461, p = .39$.

Predictors that remained in the final model were individual rating of interest-ingness, estimated profitability, and overall effort of the given HIT. According to Table 5.2, all are significant predictors of workers' decision ($p < .001$). An increase in interestingness and profitability were associated with an increased likelihood of a

---

[3] Seven participants which rated less than three HITs were discarded.
[4] The H-L test assesses the goodness-of-fit.

Table 5.2 Logistic regression analysis of 222 acceptance decisions

Predictor	β	SE β	Wald's χ^2	df	p	e^β
Constant	−.252	.708	.127	1	.722	.777
Interestingness	.246	.036	46.017	1	<.001	1.279
Profitability	53.721	14.977	12.865	1	<.001	2.14×10^{23}
Overall effort	−.05	.014	12.388	1	<.001	.951

worker accepting a HIT, whereas mean overall effort was associated with a decrease in likelihood of the worker accepting a HIT.

5.2.3   Discussion

In summary, workers have been presented a HIT and were asked to estimate its workload, and interestingness as well as to decide if they accept to perform it. The workload was determined using TLX, RTLX, and RSME measurements. Mean scores over all individual ratings were calculated for each HIT and considered as the mean workload score and the mean overall effort score for that HIT. The TLX and RTLX scores were nearly identical. Both of them highly correlated with effort considering individual ratings and mean scores for the HITs. Older participants tended to rate the effort needed for accomplishing the HITs and their interestingness less than average. Interestingness, estimated profitability, and overall effort of a HIT were predicting acceptance of HITs by workers.

It was observed that the number of participants strongly shrinks when the study consists of several jobs running one after the other. One potential reason is that next jobs got lost in the huge number of HITs available in MTurk. Besides that, it could also be the case that workers specialize themselves on a particular type of HIT (e.g. survey) and would not continue to the next job as the type of HIT will change.

This pilot study successfully showed that workload can be measured by the employed scales; RSME can be used alone in case no diagnostic information is needed. Learning from this study, the following study was conducted on a large scale.

5.3   Study 5.3: Is the Task-Selection Behavior of Workers Predictable?

This study was conducted on a larger scale and slightly differently from the Study 5.2. The aim was to create a model for predicting workers' decision on accepting or refusing HITs.

5.3.1   Method

A dataset was created by first collecting sample HITs from various types, and second assessing them by crowdworkers. In addition to the overall effort, workers determined other aspects of each HIT like interestingness, difficulty, and complexity. Figure 5.8 illustrates the procedure.

In group I, workers first answered to an Introductory job. Later, a qualified group of them were allowed to perform the HIT Rating Job with the aim of collecting five observation per each HIT from the database. For group II, workers were directly offered to perform the HIT Rating Job with the goal of obtaining ten responses per each HIT in the database. Both HIT Rating Jobs were identical. As a result, fifteen responses (as suggested by [116]) for each HIT were collected to reach a reliable score of overall measurements for concepts like overall effort, interestingness, difficulty, and complexity. Such a design was chosen to avoid several-stage study design as (1) a huge number of participants was shrinking after each stage, (2) a survey based introductory job may attract only survey professionals which can bias the data, (3) Study 5.2 showed that the overall effort highly correlates with TLX and it can be considered as a measure of workload (4) no relation between neither TLX subscales nor user characteristic and acceptance decisions were discovered. In the following, each part is described in details.

5.3.1.1   Dataset

In the first step, 400 HITs from MTurk were collected in November and December 2016 by four individuals. The main criteria for considering a HIT to be part of the

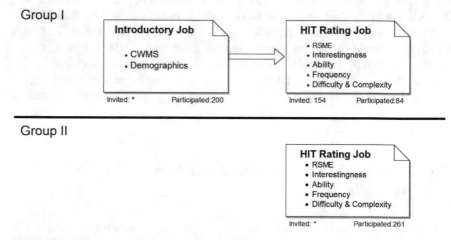

Fig. 5.8 Procedure of Study 5.3. The group I and group II provided five and ten assessments per HIT, respectively

dataset was to be able to estimate its workload from the preview page. Thus, surveys with external links and multi-page HITs were discarded. For each HIT, a screenshot from the preview page, the complete HTML source code, and the available metadata (i.e. title, description, required qualification, rewards, time allocated, available HITs) were collected. A reviewer examined the collected HITs and discarded HITs either with missing information or ones that violate other criteria. This led to 373 HITs approved to be included in the dataset. From them, 14 extensive HITs, which have a repeated pattern of questions, were selected and *manipulated* regarding their length. Consequently, for each of them, three different instances were added to the dataset namely in the original length, 50% of the original length, and 25% of the original length. For the manipulated set of HITs, it is expected that the length of the HIT is positively correlated with its estimated workload. The purpose of including the manipulated HITs in the dataset is to provide an instrument in order to evaluate the reliability of the subjectively assessed workload later. As a result, the final dataset contains 401 HITs.

In the second step, all HITs in the dataset were categorized following the schema suggested by Gadiraju et al. [30]. A crowdsourcing project was conducted in the MTurk. For each HIT from the dataset, meta-data and the screenshot were presented to three different workers, and they were asked to select the type of the presented HIT from a given list of choices (i.e. six categories proposed by Gadiraju et al. [30] and also "Do not know"). The job was available for U.S workers, who had more than 500 approved HITs with the overall approval rate of 98% or more. The same criteria were used for the rest of the crowdsourcing jobs in this study as well. The job was rewarded by $0.02. Answers from all workers were accepted. For 161 out of 401 HITs, all three workers agreed on the HIT type. The job was repeated for the remaining HITs, again three more workers determined the HIT Types. Finally, 56 HITs that workers still did not concur (less than three agreements) on their type were judged by an expert (Table 5.3).

Introductory job. For the group I, the introductory job was a prerequisite for the main study (i.e. HIT Rating job) and was available for 200 workers. It consists of two parts; demographics and the CWMS motivation questionnaires (cf. Chap. 4). The job was rewarded with $0.15 and finished within one day. Two trapping questions [82] (cf. Sect. 4.2.1.1) were employed in the CWMS questionnaire and two of the questions were repeated to check for reliability and consistency of results. Twelve respondents answered one of the trapping questions wrongly, and some level of inconsistencies were recognized in the answers of 36 participants. Although all respondents were paid, 154 workers qualified and were invited to the next job.

HIT rating job. Similar to the Study 5.2, workers were asked to assess a given HIT based on its meta-data (i.e. title, description, the allotted time, rewards, HITs available) and its screenshot (cf. Fig. 5.5). This job was identical for the group I and the group II except that five individual assessed each HIT in group I and ten in group II. Workers assessed the given HIT by answering seven questions. The first question was a binary question which asked if they are able to perform the given HIT. It was included to ensure that they are capable of evaluating the HIT. Next and similar to the Study 5.2, the RSME scale for assessing the overall effort and workload. The five remaining questions refer to interestingness, difficulty, complexity of the HIT, how often do they perform such HITs (hereafter referred to as "Frequency"), and how likely is it for them to work on the given HIT ("Acceptance").

Similar to the Study 5.2, participants rated each item (instead of overall effort) on a twenty-step bipolar scale which has been labeled accordingly (never/often, easy/hard, and simple/complex). Two simple trapping questions were included where workers should enter two numbers to proof that they checked the preview: first, the associated reward to the HIT and secondly, a two digit number overlayed on the right-bottom of the preview image.

With a 54% return rate[5] of the invited participants in group I, their HIT Ratings job finished in 22 days (i.e. five individuals judged each of 401 HITs in the dataset). The group II workers finished their job in two days (i.e. ten individuals assessed each of 401 HITs).

To encourage workers to rate more HITs, a bonus payment of $0.05 was added to the base reward of $0.10 per assignment if they performed more than 20 assignments. All participants answered the trapping questions correctly. Overall 6015 assessments for 401 HIT from 345 workers were collected. Participants judged between 1 to 234 HITs ($M = 18$, $SD = 26$).

5.3.2   Results

5.3.2.1   Data Screening

Overall 6015 responses were collected containing 15 ratings per each HIT from the sample dataset. Answers from all workers were accepted. A two steps procedure was followed to remove unreliable responses. First, each response was analyzed separately. As a result, 199 responses that have missing values and 439 responses that their workers specify they are unable to perform such a HIT were removed. In the second step, the performance of each individual worker was evaluated. To do that, a 30% trimmed mean of the overall effort (i.e. RSME) was calculated for each HIT. The trimmed mean was chosen as the arithmetic mean is sensitive to extremely deviated ratings. Given that, average absolute deviation of ratings for each worker was calculated. Workers who had on average 25 or more deviation from the mean ratings were considered to be inaccurate. All responses from them (i.e. 597) were removed. The average deviation of 25 is chosen for a cutting point as the number of suspected ratings start to exponentially raise for any cutting point below that. In addition, four HITs were also removed from the dataset as there were less than five ratings per each remaining. Overall 387 HITs remained in the dataset with 4821 valid responses given by 270 participants from them 74 individuals belong to the group I. Using all valid responses 30% trimmed mean of the overall effort was calculated for

---

[5]Invited workers were reminded two times by sending emails using the MTurkR (https://github.com/cloudyr/MTurkR) software package.

Table 5.3 Distribution of different HIT types in the final dataset (N = 387)

Type	Count	Percentage (%)
Content access (CA)	14	3.6
Content creation (CC)	129	33.3
Interpretation and analysis (IA)	96	24.8
Information finding (IF)	71	18.3
Surveys (SV)	45	11.6
Verification and validation (VV)	32	8.3

each HIT in the dataset (hereafter referred to as Mean Overall Effort). HITs were rewarded from $0 to 3.73 ($M = \$.17, SD = \$.4$) with an exception of one HIT with $29.5.

Finally, the goodness of the data cleaning procedure was evaluated based on the expected rank from the manipulated HITs in the dataset (cf. Sect. 5.3.1.1). From 14 manipulated HITs (each having three instances with different length of *short*, *medium*, and *long*), eight followed the expected order when ranking them based on the calculated mean overall effort value. For the rest, HIT instances with the medium length were ranked nearly equal or with slightly more workload than their corresponding long instances. Following the same procedure but using all responses (i.e. disregarding the data screening process) to calculate the mean overall effort, 4 out of 14 manipulated HITs follow the expected rank. In addition, the cleaned dataset has smaller SOS parameter α [43] (0.091) than the full dataset (0.139) indicating less variance in the participants' ratings. As a result, the data cleaning procedure shows a satisfactory result.

5.3.2.2   User Characteristics

In this section relationships between participants' underlying motivation and their demographic data are analyzed for the 154 participants who took the introductory job. Demographic data are reported in Fig. 5.9. The underlying motivation of participants was measured using the CWMS scale (cf. Chapter 4) in five subscales. By exploring the motivation scores using cluster analyzes three groups of participants were recognized (average silhouette width = 0.4). A TwoStep Cluster procedure of IBM SPSS is employed which automatically determines the number of clusters based on the Bayesian Information Criterion (BIC).

The motivational scores differed significantly between cluster groups, $F(8, 296) = 68.397, p < .001$; Wilk's $\lambda = .123$, partial $\eta^2 = .65$. The first group of participants (16.2%) had the lowest external motivation, but high intrinsic motivation (*autonomously* motivated). The second group (50%) had a high external motivation, but least intrinsic motivation (*Controlled* motivation). The third group (33.8%) had high scores in all four types of motivation (*Highly motivated* group).

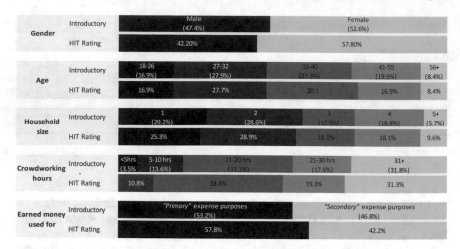

Fig. 5.9 Demographics of participants in the Introductory (N = 154), HIT Ratings Job (N = 83) from group I

84% of the autonomously motivated participants use their crowdworking income for secondary expenditures whereas 55.8% of workers in the controlled motivation group, and 67.3% of the highly motivated participants used their crowdsourcing income for primary expenses ($\chi^2(2, 154) = 18.27, p < .001$). Most of the participants in the highly motivated group are female (67.3%) whereas males are slightly dominant in the autonomously (60%) and the controlled motivation (53.2%) group ($\chi^2(2, 154) = 7.159, p = .028$). The autonomously motivated group spent less time for crowdworking (88% of them less than 20 h per week) than the group with the controlled motivation and the highly motivated ones which 53.3% and 61.5% of them spent more than 20 h per week on crowdworking respectively ($\chi^2(8, 154) = 25.972, p = .001$). 52% of the workers in the autonomously motivated group are living alone, whereas 70.1% of workers with the controlled motivation and 82.7% of the highly motivated workers are living with at least one more person in the same household.

Participants also reported the type of typical tasks that they perform in MTurk. As expected, the Survey was the most popular type of job within the participants (96.8%) followed by other HIT types: Verification and Validation (67.5%), Interpretation and Analysis (64.9%), Information Finding (58.4%), and Content Access (53.2%). The Content Creation was the least popular type of job (37.7%). The Information Finding jobs were significantly less popular with autonomously motivated workers ($\chi^2(2, 154) = 6.92, p = .031$). The Verification and Validation job type was mostly popular among the highly motivated group ($\chi^2(2, 154) = 6.023, p = .049$). Roughly half of autonomously and controlled motivated participants also took part in the HIT Rating job as well as 60% of participants from the highly motivated group.

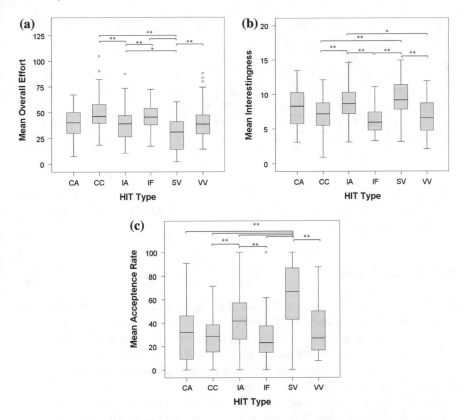

Fig. 5.10 Distribution of *Mean Overall Effort Score* (**a**), *Mean Interestingness Score* (**b**), and *Mean Acceptance Rate* (**c**) for each HIT Type in the final dataset. * $p < 0.05$, ** $p < 0.01$

5.3.2.3   Task Characteristics

Figure 5.10 illustrates the distribution of the mean overall effort score **a**, the mean interestingness score **b**, and the mean acceptance ratio **c** over HIT types.

Workload. Individual ratings of overall effort significantly correlate with difficulty ratings (Pearson's $r(4821) = .691$, $p < 0.001$) and complexity ratings (Pearson's $r(4821) = .685$, $p < 0.001$). Applying a factor analysis with the maximum likelihood extraction method and varimax rotation led to a one-factor solution which explained 81.2% of the variance. All three variables highly load on the factor ($>.78$). Its score was calculated using a regression method and employed in the following steps as a new variable (hereafter referred as Expected Workload). No significant relation between workers' characteristics and the expected workload score was discovered.

Interestingness. Workers with higher Intrinsic motivation and Identified motivation scores tended to rate tasks more interesting than the average: Pearson's $r_{intrinsic}(74) = .283$, $p = 0.014$, Pearson's $r_{identified}(74) = .260$, $p = 0.026$. In addi-

Table 5.4 Logistic regression model to predict acceptance using training dataset (N = 2721). Normalized scores were used to create the model

Predictor	β	SE β	Wald's χ^2	df	p	e^β
Constant	0.225	0.128	3.078	1	0.079	1.252
Interestingness	1.715	0.154	124.100	1	<.001	5.555
Frequency	1.395	0.121	133.938	1	<.001	4.035
Expected workload	−1.326	0.147	81.644	1	<.001	0.266
Profitability	1.819	0.308	34.773	1	<.001	6.164

tion, the mean interestingness score negatively correlates with the mean expected workload of HITs (Pearson's $r(387) = -.318, p < 0.001$). No other significant correlation was discovered.

5.3.2.4   Acceptance

For the prediction of Acceptance of a HIT, the responses from each individual worker have been analyzed, namely 4821 observations (cf. Sect. 5.3.2.1). The Acceptance score is the answer of participants to the question whether they would take a given HIT or not. This question was scored on a scale of 0–20. First, it was converted to a binary variable. Any value on this score that is below seven was labeled as a "not accepted" observation while the values above 13 were labeled as "accepted". For the rest of the values, the observations were marked unknown and removed. This created the new Acceptance variable ($N_{cases} = 3795$, Accepted = 46.2%). The dependent variables were from meta-data i.e. Rewards, HITs available, and subjective assessments of workers i.e. Expected Workload, Interestingness, Frequency (how often worker performs this kind of HITs), and the ratio of the reward to the Expected Workload (hereafter profitability). For subjective variables both worker's specific ratings and mean score over ratings of all workers for that specific HIT were considered. As user characteristics were not available for all cases, they were not included in the primary model. A Binomial Logistic Regression model was built employing the training dataset that includes 72% of observations (i.e. 2721). The logistic regression model delivered an accuracy of 88.8% (sensitivity: 85.5%, specificity: 91.1%), and was found to be significant with $\chi^2(8) = 34.32, p < .001$ and explained 72.5% (Nagelkerke R^2) of variance. Besides the predictors mentioned in Table 5.4, the rest of the dependent variables were found to be insignificant and were not used.

Later, the model was evaluated with the test dataset (i.e. remaining 1074 observations). The test results accuracy of 89.54% (sensitivity: 88.47%, specificity: 90.82%). As a result the model is accepted to be a predictor of user's decision on taking the task.

The effect of users' underlying motivation was evaluated using 1467 observations which were provided by participants in group I. From them, 363 ratings collected from 12 participants which basically were autonomously motivation, 551 ratings from 27 workers who were highly motivated, and 553 ratings from 30 workers who belonged to the controlled motivation group. Note that not all participants rated the same set of HITs. Effects have been evaluated in three steps. First, a logistic regression model was created given the previously used predictors and scores of the CWMS's five subscales. The employed variables in the final equation were selected using the Forward Selection (Likelihood Ratio) method. From the motivation subscales, just the score of Identified Motivation remained in the model. Results showed that for each point increase in Identified Motivation, the odds of accepting a given HIT increases by 0.27 (Wald's $\chi^2(1) = 7.425$, $p = .006$). This model predicted users' decisions with an accuracy of 88.5% (sensitivity: 83.7%, specificity: 91.8%). However as the previous model also achieved an identical result, the effect of the new variable was not investigated further.

In a second attempt, the motivation group was employed as a predictor in addition to the accepted model. The same procedure as in the first trial was followed. Results showed that when all other factors are similar, the odds of taking a HIT by a person who belongs to the autonomously motivated group is half of the odds of a person who belongs to the highly motivated group ($e^\beta = .499$, Wald's $\chi^2(1) = 11.654$, $p = .001$).

In the last attempt, for each group of participants, a separate logistic regression model was fitted with the same set of predictors used in the final model (cf. Table 5.4). Figure 5.11 illustrates the 95% confidence intervals of odds ratios for each predictor. Surprisingly, Interestingness tends to be more important for the highly motivated group than the autonomously motivated ones. For the later one, the probability of accepting a HIT that is similar to what they typically perform is higher than the other groups. However, expected workload does not influence their decisions as much as the other groups (especially highly motivated ones). The highly motivated group considered the profit (ratio of reward to workload) more than the others.

Further investigations using mediation analysis and all observations showed that there is a significant indirect effect of Interestingness on Acceptance decision through Frequency of taking similar HITs, $\beta = .73$ BCa CI [.61, .85] (Sobel test $z = 12.41$, $p < 0.001$). In other words, workers are interested in the HITs that they perform typically (Fig. 5.12). Considering the Fig. 5.11a, Interestingness also has an indirect effect on Acceptance which is not considered there.

5.3.3 Discussion

A logistic regression model for predicting the Acceptance of a given HIT based on users' ratings of Interestingness, the frequency of performing similar HITs, its expected workload, and its profit (ratio of reward to the expected workload) was created. The model could predict users' decision with an accuracy of 89.54%. Considering the odds ratio, Profit has the highest effect, followed by Interestingness, Frequency, and Expected workload. However, their difference was not significant. Considering the indirect effect of Interestingness on the Acceptance through the

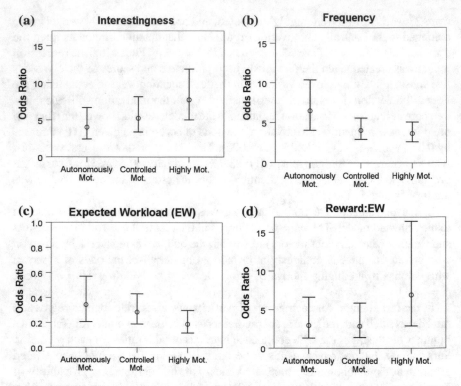

Fig. 5.11 The odds ratio of predictors fitted to ratings collected from participants in different motivation groups

Fig. 5.12 Final acceptance prediction model. ***
$p < .001$, a: Sum of direct and indirect effects

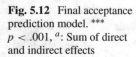
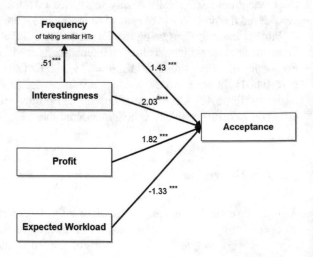

mediation of Frequency, the Interestingness can move up to first place on the list. It was also shown that depending on the underlying motivation of participants, the effect of each factor might change. Highly motivated participants look for profit and follow their interest. They are less focused on a particular type of task than the others. Autonomously motivated ones, mostly specialize on tasks and they choose tasks based on their interest. A task being associated with slightly higher workload does not degrade their motivation as much as it does for other workers.

Note that individual ratings were kept in the final model and mean scores were removed by the Forward Selection (Likelihood Ratio) method. Applying the mean scores of Interestingness, Frequency, and Expected Workload leads to a model which can predict the Acceptance with an accuracy of 69.9%. Aforementioned indicates that individual ratings or precisely the deviation of an individual's rating from average score contains valuable information which increases the accuracy of prediction up to 20%. In the case of Expected Workload, the deviation can indicate one's skill level and experience in that area, whereas differences in Interestingness ratings shows the personal preference.

One drawback of this study is that the effect of the requester is completely neglected. The HIT's requester was removed from tasks' meta-data although it was in third place on the rank of important factors for workers (cf. Study 5.1). In addition, the effect of rewards could not be analyzed in detail as it was a static amount of compensation permanently attached to each HIT. Last but not least, just the U.S crowd workers were investigated in this study.

The next study is conducted to examine the effect of requester, rewards and the important factors found in the current study on users decision in a different population.

5.4   Study 5.4: What is the Relative Importance of Influencing Factors on Workers Decision?

With the goal to further understand the crowdworkers' preference for microtasks an Adaptive Choice-Based Conjoint (ACBC) analysis was conducted. The Conjoint Analysis (CA) is a popular technique orginated in mathematical psychology and used in different domains like marketing research to determine what features a product should have and how it should be priced [89]. The general idea of CA is that one evaluates the overall desirability of a complex product or service based on a function of the value of its separate components [89]. Each component of the product is considered as an *attribute* which can have one or more *levels*. Mostly all attributes of a product are important in their own sense, but in a conjoint survey respondents must trade off different aspects of the product (as in real life). As a result the relative importance of each attribute can be inferred

In a full factorial design, all combinations of attributes and levels are created which leads to a full set of possible product *concepts*. Then respondents are asked to rank or choose (i.e. choice-based CA or CBC) among the presented set of concepts. By systematically varying different features and observing how respondents react to the resulting product concepts, one can statistically infer (typically using linear regression) an individual value (i.e. part-worths) that respondents subconsciously consider for that attribute [89]. As the number of attributes and levels increases, the number of possible concepts resulting from full factorial design exponentially increases. In such a case, mostly the Adaptive CBC (ACBC) should be employed.

The ACBC questionnaire consists of three parts. First, *Build Your Own* (BYO) section, in which participants create their own ideal product (i.e. HIT) based on available attributes and levels. Rationally one would prefer to take a best combination i.e. least workload, most profit, and highly interesting. Therefore, the *better* levels are associated to decrease of HIT's reward (or in case of product increase of cost). The goal is to find the most relevant concept space for each participant. In addition, it is a quick and easy anchoring session for participants in which they are educated about all available attributes and levels [95]. Next, in the *Screening Process*, a set of relevant HITs are created based on what participant specified in the BYO section. Participants are ask to specify if each of given HITs can be included in their consideration set. Here the goal is to investigate non-compensatory rules i.e. must-haves or unacceptables levels of attributes to create a manageable consideration set [38]. The last step is a *Choice Tasks Tournament*. Similar to CBC, three or more HITs are presented to the participants in each page and they are asked to choose their ideal HIT that they would perform. However, just concepts which are close to the consideration set of participants will be used. As a result, the number of questions that participants need to answer strongly reduces. Several studies have shown that the ACBC is as accurate as the traditional CBC [14]. In this study Sawtooth SSI Web[6]

5.4.1 Method

An ACBC questionnaire was created to assess the relative importance of HIT's attributes and attributes' levels for crowdworkers. Figure 5.13 illustrates the steps participants took in the conjoint survey.

5.4.1.1 Attributes and Levels

Based on the findings of aforementioned studies and also the state-of-the-art, the following attributes and levels were considered in the study. Later, one level from each attribute will be used to automatically assemble sample HITs which will be assessed by workers.

---

[6]http://www.sawtoothsoftware.com/products/online-surveys. (v 8.2.4) is used for creating and analyzing the ACBC test.

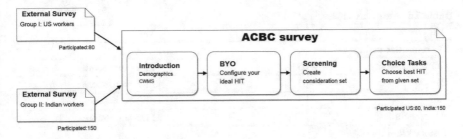

Fig. 5.13 Procedure of Study 5.4

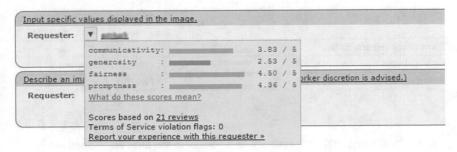

Fig. 5.14 Example of requesters' ratings in MTurk using TO browsers extension

Requesters' ratings. As it was found in Study 5.1, requester's ratings is the third factor in the rank for crowdworkers to decide on taking a HIT or not. However, it was neglected in Study 5.2 and 5.3. Here, requester's ratings provided by Turkopticon (cf. Sect. 1.1.1.1) were used.

Turkopticon provided five-point ratings for the following four dimensions and present the aggregated ratings of each requester on MTurk listings (cf. Fig. 5.14, [56]):

• Communicativity: How responsive the requester has been to communications or concerns workers have raised.
• Generosity: How well the requester has paid for the amount of time their HITs take.
• Fairness: How fair the requester has been in approving or rejecting works.
• Promptness: How promptly the requester has approved works and paid.

Task's meta-data. Difallah et al. [21] used data from MTurk Tracker[7] in the time range from June to October 2014 (hourly observations) to predict batch's *throughput*, which they defined as *"the number of HITs that will be completed for a given batch within the next time frame of 1 h"*. This can be considered as a measure on how preferable a HIT is for workers. Their results showed that the number of tasks left in a HIT group (*HITs Available*) and how recently the HIT is created (*Created Time*) are two key features having a statistically significant impact on the throughput score with

---

[7]Since 2010 collects data about HITs published on MTurk. http://mturk-tracker.com/.

Table 5.5 Attributes and levels of HITs

Attribute	Level	Value
TO requester rating		
Communicativity	5	Five-point score
Pay	5	Five-point score
Fair	5	Five-point score
Promptness	5	Five-point score
Tasks meta-data		
HITs available	5	1, <10, 10 − 100, 101 − 500, 500+
Created on	5	Just now, Last 5 min, Last hour, Today, This week or earlier
Tasks perceived attributes		
Expected workload	3	Perceived from given HIT
Interestingness	3	Perceived from given HIT

$p < 0.05$ and $p < 0.001$, respectively. In addition to the aforementioned attributes, *Reward* was also included in the attribute list based on the results of Study 5.1 and 5.3.

Task's perceived attributes. Based on the results from Study 5.3, Expected Workload and a task's Interestingness were two task attributes that were considered in this study. As both of them are perceived concepts, we decided to provide sample HITs with different levels of each attribute rather than statements like "high expect workload", or "very interesting". Three levels for each attribute were considered which leads to nine combinations in a full matrix design. Using the dataset of Study 5.3, nine HITs were selected which cover all combinations between three levels of Expected Workload (easy, medium, and hard) and three levels of Interestingness (low, medium, and high). Note that although Profit and Frequency were found out to significantly predict workers decision in Study 5.3, they were not included as attributes in this study. Profit was redundant as both Reward and Expected Workload were included. On that point of time, we could not find any practical way to include Frequency as an attribute in a way that participants perceived its value and other attributes within the study.

Attributes and their corresponding levels used in this study are summarized in Table 5.5 and Fig. 5.15 represents a choice tasks tournament in ACBC interview.

5.4.1.2   Study Design

Two survey-link jobs were created in MTurk to recruit participants for this study. The job contained a short description of the study and a link to the external web page which served the ACBC questionnaire. One job was available for the U.S based and the other for the Indian based workers. In addition, workers should have more than

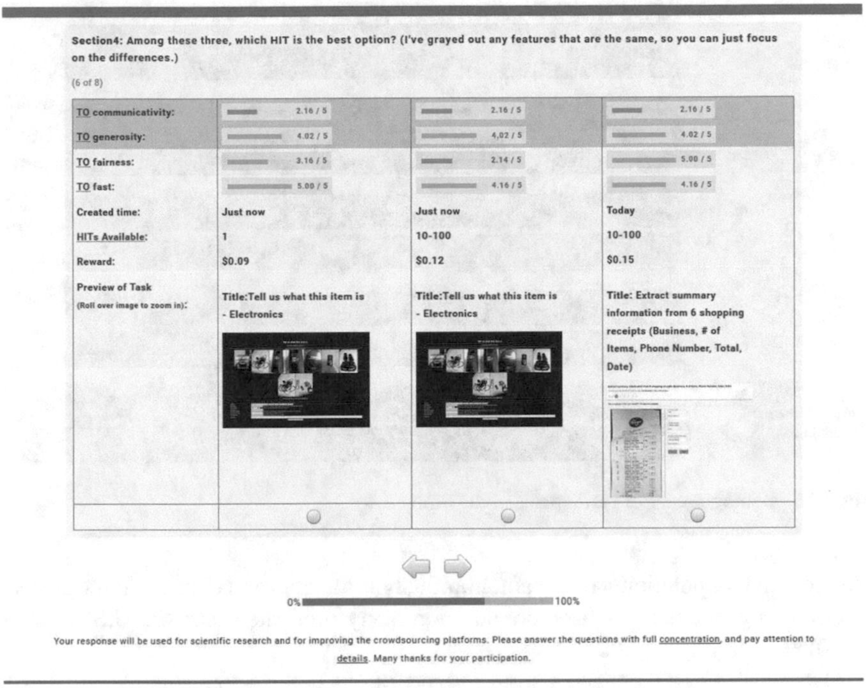

Fig. 5.15 Choice tasks tournament of ACBC interview. Participants could view the full-size screenshot of the given HIT by rolling over the thumbnail or clicking on that

98% approval rate and 500 or more accepted HITs to be able to participate in the study. Overall 80 U.S based workers and 150 Indian based workers participated in the study. The job was compensated with $.75. All responses were accepted.

5.4.2   Results

Two trapping questions previously included in the CWMS questionnaire were used for the reliability check. Results from 78 U.S workers and 114 Indian workers were considered reliable. Demographic data of participants are illustrated in Fig. 5.16. The population of Indian participants in this study concurred with the findings in Chap. 3. They are mainly male, younger, and living with more people in the same household than their U.S counterparts. More than half of Indian participants stated to use their crowdworking's income for covering their primary expenses. Based on the Relative Autonomy Index (RAI)[8] two clusters of participants were discov-

---

[8]$RAI = 2 \times Intrinsic + Identified - Introjected - 2 \times External$.

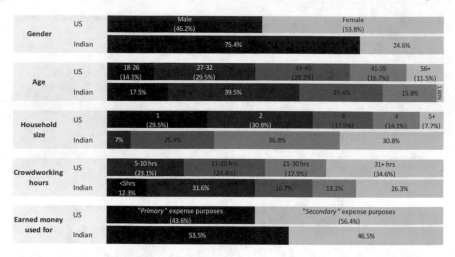

Fig. 5.16 Demographics of participants in the Study 5.4

ered for each population namely autonomously motivated and controlled motivated. The Indian participants were more autonomously motivated than the U.S workers $t(190) = -4.724, p < .001$.

Overall 3323 HIT concepts were judged by 78 U.S respondents ($M = 42.6$). Associated rewards to concepts were between \$0.02 to 0.58 ($M = \$0.24, SD = \$0.1$). The 114 Indian crowdworkers assessed 4690 HIT concepts ($M = 41.1$) which were rewarded between \$0.02 to 0.36 ($M = \$0.16, SD = \$0.06$). Note that the SSI Web software calculated the reward of each HIT concept dynamically based on a sum of the costs associated with individual levels which were used in that concept. The cost of an attribute's level was assigned during the study design. Also, the software dynamically varies the reward up to 30% above the summed price and down to 50% below the summed price based on the given responses. Individual part-worth utilities and relative importance of each attribute were estimated for every participant using the Hierarchical Bayes (HB) models.[9] The software estimates an HB model using a Monte Carlo Markov Chain algorithm [88].

Relative importances. It shows how much difference each attribute could make in the total utility of a HIT. Estimated mean values and 95% CIs are illustrated in Fig. 5.17. There was a statistically significant difference in the importance of attributes based on the population, $F(8, 1728) = 13.163, p < .001$; Wilk's $\lambda = .635$, partial $\eta^2 = .365$. Overall, Rewards was the most important attribute for both U.S and Indian crowdworkers. From the task-related attributes, Expected Workload and Interestingness were ranked above other attributes for the U.S workers. However, the Indian workers considered that the number of available HITs and creation time were as important as the Interestingness of the HIT. Those attributes were significantly more important for Indian workers than the U.S workers. From requester's attributes,

---

[9]HB is the recommended method provided by the SSI Web software.

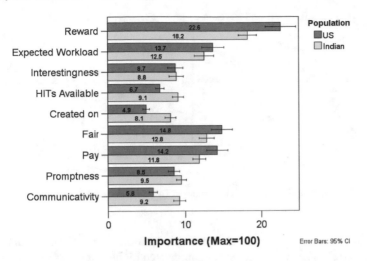

Fig. 5.17 The relative importance of HIT's attributes for the U.S (N = 78) and India (N = 114) workers

both groups agreed that Fairness and Generosity are the most important ratings to be considered. Communicativity was significantly more appreciated by the Indian workers than by their U.S counterparts. Rather than Interestingness, Expected Workload and Requester's Promptness, the importance of the other attributes significantly differs between the U.S and the Indian worker groups. No significant difference was discovered between the importance of rewards for Indian workers with the controlled underlying motivation and the U.S workers.

Part-worth utilities. The part-worth utilities can only be compared within each attribute and should be treated as an interval data type. Within an attribute, the direction of the part-worth utility scores indicates the preferences of the respondents (i.e. levels with positive scores are preferred over those with negative ones). A higher value of part-worth utility indicate a stronger preference [86]. The part-worth utilities for two task-related attributes and two requester-related attributes are reported in Fig. 5.18.

The U.S workers preferred tasks with medium to high Interestingness whereas Indian participants preferred the less interesting tasks. There could be two reasons associated to that. Either the selected tasks were not perceived as interesting as they were expected to be for the Indian workers (as U.S crowdworkers rated the Interestingness of them in the Study 5.3), or – as higher Interestingness were associated with fewer rewards- they more preferred the gain obtained from less interesting HITs. Both populations preferred HITs with an easy to medium expected workload. Moreover, autonomously motivated workers preferred HITs with medium workload equally or slightly more than the easier HITs. However, in the case of hard HITs, the preference of autonomously motivated workers dropped stronger than the workers with controlled motivation.

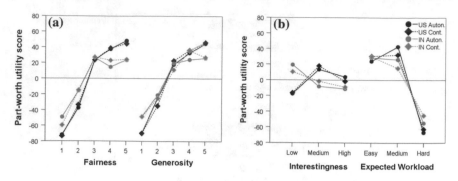

Fig. 5.18 Part-worth utilities for selected requester (**a**) and task-related (**b**) attributes

For both fairness and generosity ratings, requesters with three or more points were preferred. The U.S workers still appreciated higher fairness score but not their Indian counterparts. The U.S workers, as well as the Indian workers with underlying controlled motivation, preferred three or more points for the generosity ratings.

5.4.3 Discussion

The conducted ACBC study showed that different groups of crowdworkers evaluate HIT's attributes differently. The associated reward was the most important criterium for workers when choosing between HITs. From tasks attributes, expected workload and interestingness were in the next places which validate the results of Study 5.3. However, for the Indian workers, other task attributes were as important as the interestingness which indicates that the model should be customized for the different groups of crowdworkers. Additionally, Indian workers prefer the HITs with low interestingness. It might be due to the fact that the Indian participants perceived the given HITs not as interesting as their U.S counterparts who rated those HITs on the first occasion in the Study 5.3. Note that one drawback of ACBC studies is that estimated importances of attributes depend on the particular attribute levels chosen for the study [89]. A narrower range of levels for an attribute can make that attribute less important. The nine sample HITs employed in this study had an interestingness score range from 3.5 to 11.1 on a 20-point scale.

Among requester's ratings, the fairness, and the generosity were the most important attributes. They have been shown to be as important as Expected Workload. Workers trust to the requesters with adequate profiles scores to properly pay for the amount of time their HITs take, and fairly judge the quality of their answers. Therefore, other factors rather than the tasks-related attributes influence the workers' decision on choosing their HITs. Consequently, the model proposed in the Study 5.3 can be extended to include the fairness and generosity ratings of the requesters.

5.5   Study 5.5: Is the Expected Workload Predictable?

The model built in Study 5.3 can predict whether an individual accepts a HIT or not given personal assessments of HIT's interestingness, expected workload, and the frequency performing similar HITs. It was shown that although using the mean interestingness and the mean expected workload scores lead to less accurate prediction, still such a model can predict workers' decision with an accuracy of 69.9%. In this study, a model for predicting the mean expected workload of HITs from its design are provided. Such a model can then be used to enhance the acceptance prediction model. In addition, a growing body of literature has addressed ethics including fair payments in current crowdsourcing microtask platforms [4, 27, 63, 87, 97, 104]. Being able to measure workload corresponding to a microtask is the initial step towards fair crowdworking.

5.5.1   Method

The dataset created in the Study 5.3 was employed (cf. Sect. 5.3.1.1). Consequently, for each HIT in the dataset, its meta-data, a screenshot from the preview page, and its complete HTML code in addition to the mean expected workload score were available. Note that the manipulated HITs were removed as their HTML codes were not corresponding to the screenshot and the expected workload ratings. Consequently, 345 HITs remained in the dataset. Relevant features from each HIT were extracted and different models for predicting the mean expected workload score of HITs were created and evaluated. In the following, each step is described in detail.

5.5.1.1   Feature Extraction

Three groups of features were extracted and hypothesized to be relevant for predicting the HIT's mean expected workload. *Syntax* and *Semantic* based features were extracted by parsing corresponding HTML code and analyzing the text content respectively. The *Visual Features* were extracted from the HIT's preview image and mostly indicate how dense the text content was presented in the task.

Syntax based features. The syntax analysis aims to obtain information about the workload of the task based on the quantity and type of Standard Generalized Markup Language (SGML) elements appear in the HTML code. The SGML elements represented by *tags* indicate the type and amount of information that needs to be processed by the worker and specify which inputs are required to finish the task.

The number of <img>, <audio> or <video> tags indicates how much and what type of information the workers have to process before they can provide required answers. A more intimidating tag that frequently appears in the HTML code is the <a> tag, which implements links. A potential workload associated with an external link is unclear as the code does not reveal the content behind the link. The <select> and <input type = "radio"> tags indicate single option, and <input type = "checkbox"> tag refers to multiple option questions. Here smaller workload due to the given predefined answers are expected, but workers need to make a decision. Another common input type among the HITs is the <input type = "text"> which lets the worker freely type in their text input. The associated workload to this tag is unclear without a semantic analysis as the tag alone does not reveal how much effort is required to produce the unknown required amount of text.

Furthermore, the tags do not only provide information about the task itself but also on how it is presented. One can further identify the size, style, and color of the font.

Overall 28 features were extracted by counting different types of tags in the syntax analysis (See A.1.1 for the complete list).

Semantic based features. In the semantic analysis, the syntax was completely disregarded, and only the text seen on the web page by the worker was investigated. The Standford CoreNLP Toolkit [71] was used for parsing the text, creating a parse-tree and extracting semantic features. Statistical features like the number of words and sentences, average sentence length, and the average number of subclauses were extracted. It was hypothesized that long and complicated instructions would increase the effort associated with the HIT and consequently workload of the HIT. Employed vocabularies were also investigated. The range of vocabulary used in the text may be an indicator of how easy the task is to comprehend. A catalog of "difficult" vocabularies was created by assigning a "rareness" score to all word stems used in the HITs dataset. Later the words appeared in a given HIT were checked against the catalog, and the number of words with rare stems were calculated. Overall six semantic based features were extracted for each HIT in the dataset (See A.1.2 for the complete list).

Visual features. Visual features were extracted by processing the screenshot captured from the preview page of each HIT (cf. Sect. 5.3.1.1). Visual features are desired that might influence workers' estimation of workload while looking at the preview of a particular HIT. Web designers follow different theories and practices to design a visually appealing web page that can convey the right information with minimal difficulty to the user. While a large number of quantities differentiate one HIT from another (e.g. the style, format, font), here representations of two aspects, namely text and media inside the HIT, were considered. In the domain of Image Processing, Saliency Detection is a common method for extracting important objects from an image, or highlighting regions that stand out from other parts [67]. The saliency level itself is a measure of the importance of parts of the image that will be visually captured by the user. Previous research showed that saliency correlates well with spending human attention as a resource [102]. Hence it might also relate to expected workload. Usually, quantities like color gradient and edge frequency in the image are used in saliency detection. In this study, the saliency map calculated by the Saliency Toolbox [112] were employed.

Furthermore, as workers assess a HIT by browsing the preview page, they most probably read parts of the visible text to learn what is expected from them. Therefore, the representation of text related features (extracted from the image) was also considered to be relevant. The OCR (Optical Character Recognition) operation provided in the Matlab software package[10] was used to extract text from the preview screenshot. Then, the image was binarized based on presence and absence of text in the image. Finally, the saliency map of the image and the text regions were merged to a new binary image called composite image. The procedure is illustrated in Fig. 5.19. The composite image was divided into smaller windows, and 13 visual features were extracted. In the following selected ones are described (See A.1.3 for the complete list):

- **Text density**: Ratio of the white pixels (text in the images) to the size of the window and used to calculate further feature.
- **Threshold**: Mean (over all windows) text density in the image.
- **Quartiles**: Using an Inverse Cumulative Distribution function over an image, the 25, 50, 75, 95, 98, and 100% quartiles of text density were calculated. The 25% quartile, indicates the text density value that 25% of windows have less text density than or equal to that. They were calculated by just arranging the text density values in ascending order.

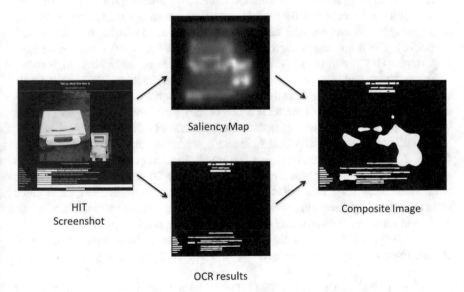

Fig. 5.19 Visual features extracted from the screenshot image of the preview page. The composite image is a result of morphological OR between the saliency map and the OCR results

---

[10]Release 7.1.0 was employed.

- **Count of dense windows**: The number of *dense windows*. They are windows with text density higher than a specific threshold.
- **Largest connected region**: Connected regions were created using a 8-connectivity neighborhood over the dense windows. This quantity is stored as number of pixels and also as number of windows.

5.5.2 Results

The dataset including 345 HITs was used to predict the mean expected workload for each HIT based on the extracted features. The three feature sets mentioned above were used to train and test two prediction models one based on the Linear Regression and the other based on the Random Forest Regressor.[11] Many of the extracted syntactic features can be considered as noise, because of their low observation frequency in the HITs. Therefore, the model that used all available features exhibited a high standard deviation and can be considered as overfitted. The Automatic Linear Modeling in IBM SPSS (version 24) was employed to build a linear regression model. The full set of features was used in the forward stepwise [24] method to automatically select predictors. Marginal contribution of predictors, entered or removed from the model, were evaluated using Akaikes Information Criterion Corrected (AICC) [51]. In addition, stability and accuracy of the model were enhanced using the Bagging (Bootstrap aggregating [8]) technique. That leads to an ensemble model with adjusted $R^2 = 0.55$ (Fig. 5.20). The following predictors were used in the final model ordered by their importance: HIT Type, 100, 98, 75, and 95% quartiles, Subclauses, Reward, Number of unique stems, and Number of words (See A.2 for the exact model). The ensemble linear regression model predicts the mean expected workload with a mean absolute error of .3 ($SD = .24$). This model is a good estimator of the expected workload as 72% of predictions were inside the 95% confidence intervals (Table 5.6).

A similar process was repeated to create a prediction model using the random forests. To avoid overfitting, recursive feature elimination [15] in combination with a random forest regressor was used to determine the 20 most important features, which were then used as a reduced feature set. From them, another 12 features were eliminated by discussing each features importance individually and testing the explained variance. In an arbitrary order, the remaining and best performing features were: HIT Type, Reward, Number of Words, Number of Subclauses, Number of HTML tags: $< img >$, $< textarea >$, $< Input >$ (type: radio), and the 100% quartiles.

With a mean absolute error of .29 ($SD = .51$), the resulting random forest model proved as a good estimator for the mean expected workload. The predicted values were inside the 95% confidence intervals in 74.78% of HITs (Table 5.6). Examining

---

[11]This model was created during the student project, and reported for the sake of completeness [103].

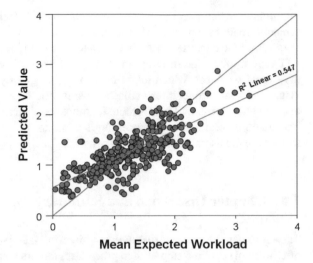

Fig. 5.20 Mean expected workload predict results using linear regression

the importance of features shows that reward is contributing the most information when predicting the expected workload. The semantic features number of words and number of subclauses, as well as HIT's type have a similar significance and hold relevant information. The remaining features ranked fairly low, which might be due to their low variance.

Last but not least, the predicted mean expected workload was used in the model provided in the Study 5.3 for predicting the users' decision on accepting a HIT. Using the individual ratings of interestingness, frequency and the predicted mean expected workload and profit, the model achieved an accuracy of 77.21% (sensitivity : 77.71%, specificity: 76.77%) using the complete dataset i.e. 12.33% decrease of the accuracy of acceptance prediction.

5.5.3   Discussion

Results of this study showed that the mean expected workload score could be predicted given the task design. This value corresponds to how an average crowd worker estimates the workload associated with a given HIT. Type of the HIT is the most

Table 5.6 Results of mean expected workload prediction with different models

Prediction model	r	MAE (std)	Overlapping 95% CI
Base model	$-^a$	.47 (.35)	187
Random forest	0.75	.29 (.51)	258
Linear regression	0.74	.3 (.24)	248

[a]Cannot be computed because base model is constant

important predictor in the linear regression model, following by visual features, semantic features, and reward. The fact that reward is found to be a major feature seems intuitive because it usually correlates with required effort. The importance of HIT type implies that different models could be used per HIT category to predict the expect workload which might lead to higher quality. Although the introduced visual and semantic features outpace the syntactic ones, they should be considered as elementary features. There is room for more sophisticated research using advanced image and natural language processing techniques on relevant visual and semantic features to the users' estimation of workload.

5.6 Chapter Discussion and Summary

The above chapter presents five studies addressing factors influencing the task choice of crowdworkers. In a step by step procedure, factors were determined, instruments for measuring them were investigated, their influences were analyzed within different worker groups, and finally their prediction based on the microtask design was investigated.

First, in a qualitative study, a list of important factors was collected from crowd-workers. For a set of HITs, participants specified why they do (not) accept to perform each of them. Three major categories of factors were discovered namely payout, requester reputation, and tasks-specific attributes. The next study focused on instruments for measuring task attributes, namely interestingness and workload. Results showed that overall effort measured by one-dimensional RSME scale adequately estimates workload of a microtask, and using a multidimensional diagnostic scale like NASA-TLX is not necessary. In the third study, a large dataset of actual MTurk HITs was collected and used for modeling workers' HIT acceptance. The result is a logistic regression model that predicts workers' HIT acceptance based on individual assessment of HIT's interestingness, expected workload, frequency of performing similar HITs and the profit (i.e. ratio of reward to the expected workload). In 89.54% of cases, the model could correctly predict workers' decisions. Using the mean scores instead of individual ratings dropped the accuracy to 69.9%. In the follow-up study, a linear regression model and a random forest model were created for predicting the mean expected workload of a HIT given its design. The linear regression model used HIT type and associated reward in addition to visual and semantic features for predicting the mean expected workload. Predicted values largely correlate with the original scores ($r = .74$). Next, the predicted expected workload was employed in the acceptance model replacing the personal rating of the expected workload. The model predicts users decisions with the accuracy of 77.21%.

Last but not least, the influence of requester ratings and task-related factors was studied in an ACBC study in different worker groups. Results revealed two important aspects. First, requesters' fairness and generosity are important influencing attribute on workers' decision. Second, the importance of attributes significantly differs between worker groups. Indian workers considered rewards less important (although still most important attribute) than what their U.S counterparts consider. In contrast, they value the number of available HITs (from one HIT group) and creation time of the HIT as important as its interestingness. However, the most important attributes were ranked equally. It was also shown that the underlying motivation of workers influences the coefficient of predictors in the acceptance model (however it was not significant). It could be due to the prerequisite of a trustworthy profile statistic (98% or more approval rate in 500 more HITs) for participating in our studies. As a result, separate fits of the acceptance model for the different worker groups are recommended.

The provided acceptance model matches to the SDT motivation spectrum. Amotivation refers to the absence of motivation which corresponds to the situation when a worker refuses to perform a HIT i.e. insufficient motive. Profit refers to the external regulation. High profit associated with a task leads to external (controlled) motivation. Frequency refers to the identified regulation in the CWMS or the integrated regulation in SDT,[12] namely "Because this job is a part of my life". Last but not least, interestingness refers to the intrinsic motivation of the task. Introjected regulation was not directly observed in the proposed acceptance model.

Further investigation are necessary to enhance the acceptance model. As reported, replacing the individual rating of expected workload with the predicted mean expected workload leads to 12.33% decrease of the accuracy of acceptance prediction. The difference shows that individual ratings include other important aspects like a personal level of knowledge, skills, and experiences which should be considered in the acceptance model. A worker with a higher writing skill might estimate workload of a summarizing task significantly less than a worker with a lower skill. In future works, relevant indicators from workers' profiles (e.g. type of the tasks previously performed) should be included in the acceptance model to enhance its prediction accuracy when mean scores are employed. Moreover, a model for predicting the interestingness can replace its individual ratings in the acceptance model as well. The same applies here, as different worker groups perceived HITs' interestingness differently. First signs have been observed in the ACBC study: Indian participants have preferred the HITs which scored less interesting by U.S workers than the one scored higher.

The proposed linear regression model for predicting the mean expected workload can be enhanced by employing more sophisticated visual and semantic features. In case that a reward (which currently is a predictor) can be replaced by other features, the enhanced model can be used to estimate fair reward associated to a task. It is also interesting to investigate how an expected workload differs from the real workload.

---

[12]Integrated and identified subscales were merged in CWMS questionnaire (cf. Sect. 4.2.2.2).

One limitation of this chapter is that in all studies, workers were stating their *willingness* to take a given HIT rather than actually taking it. Although there is no motivation for them to provide an unreal statement and different methods for removing unreliable data were used, individuals still might behave inconsistently, when they do not need to support their choices with real commitments [79]. A similar effect is known as *hypothetical bias* when participants are asked to report their monetary willingness-to-pay for a good (see [68, 79]). There is no broadly accepted theory of respondent behavior that explains hypothetical bias [68]. To reduce this effect, different approaches, mainly applied to the study design, are proposed in the literature such as using choice-based elicitation mechanism [79] and reducing social desirability bias [68]. The studies reported in this chapter might be influenced by the hypothetical bias in a way that in reality workers may not take all HITs that they stated. Such an effect and its influences need to be considered in future studies.

Chapter 6
What Influences Workers' Performance?

The inadequacy of collected data through crowdsourcing microtask jobs was addressed numerously. Different reasons were given, namely task related (ill-designed jobs like given instruction may be misunderstood), worker related (participants who do not focus enough, share their attention with a parallel activity and do not work as instructed or try to maximize their monetary benefit with minimum effort), and environment related (workers may be interrupted or use inadequate equipment) [77, 82, 83, 93]. To reduce their impact, different quality control approaches were proposed such as using gold standards, majority voting, and behavioral logging to evaluate reliability of the collected data in post-processing [17, 31]. In the context of QoE, researchers have used additional methods like content questions and consistency tests [46, 47].

In this chapter the influence of workers' underlying motivation and task's design on the *performance* of crowdworkers are investigated. Different performance indicators are considered depending on the study. Mostly, they refer either to the reliability of responses or participation of workers. Based on the taxonomy provided in the Chap. 2, the type, and amount of worker's motivation influences their performance. Furthermore, task design indirectly influences the motivation and performance of workers. These hypothesized relationships are investigated in this chapter.

6.1 Motivation

6.1.1 Survey Task

Data collected in the Study 4.1 were used to evaluate the hypothesized relationship between motivation types and worker's *participation* and their *quality of responses*. As explained in the Sect. 4.2, U.S workers with different HIT approval rates were

© Springer International Publishing AG 2018
B. Naderi, *Motivation of Workers on Microtask Crowdsourcing Platforms*, T-Labs
Series in Telecommunication Services, https://doi.org/10.1007/978-3-319-72700-4_6

Table 6.1 Correlations between motivation types and outcomes (only the reliable data are used)

Scale	Worker participation	Inconsistency score (inverse of reliability of response in survey)	Overall HIT approval rate group[a]
Intrinsic	.170[c]	−.335[c]	.003
Identified	.319[c]	−.162[c]	.148[b]
Introjected	−.075	−.138[b]	−.098
External	.079	−.07	.126[b]
Amotivation	.029	.164[c]	.033

[a] A Spearman's rank-order correlation
[b] Correlation is significant at the .05 level (2-tailed)
[c] Correlation is significant at the .01 level (2-tailed)

invited to answer a survey including the CWMS questionnaire. The survey also included items to measure workers' participation which refers to the amount of activities that workers perform in the MTurk. It was measured on three items (i.e. *How much time do you usually spend per week on MTurk?*, *How many times do you usually visit the MTurk website?*, *How many HITs do you usually complete per week in MTurk?*) which were either answered in free text input format (later binned to 7 bins) or rated on a predefined 7 point scale (i.e. *One time per month or less* to *More than once a day*). All items were averaged to create the *worker participation score*, with a good Cronbach's α of 0.822.

Two criteria were used as an indicator of the quality of response. First, the Inconsistency Score (cf. Sect. 4.2.1.1) as an indicator of the reliability of a response in the given task, i.e. filling the questionnaire (the higher the IS, the more unreliable the response).[1] Second, the workers' overall approval-rate group as an indicator of the reliability of their responses in the long-term (over all requesters and types of jobs). Five different groups were available as five survey jobs were launched each with a different range of approval rate as a required qualification.

Correlations between the motivation scores of each subscale and worker's performance, observed in the Study 4.1, are reported in Table 6.1. Results follow the hypothesized pattern between motivation types and the two outcomes, i.e. the worker participation and the task response reliability. Autonomous motivation is positively related to the participation; and a higher degree of autonomous motivation leads to a higher reliability of the answers. However, the Spearman's rank-order correlation between motivation subscales and the long-term approval rate were small. This is possibly due to motivation being a dynamic construct changing over time and dependent on additional characteristics such as the task design.

---

[1] Here, the distribution of IS for approved responses is considered. The responses with unacceptable IS (i.e. higher than the cut-off threshold) were already removed in the reliability check phase and not included in any analyses.

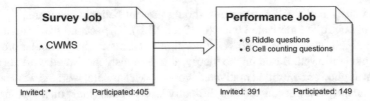

Fig. 6.1   Procedure of the Study 6.1

6.1.2   Study 6.1: Algorithmic and Heuristic Tasks

6.1.2.1   Method

Participants from the Study 4.1 were invited to a follow-up study (Fig. 6.1). The study was limited to one job which consisted of two parts. In the first part, participants should find answers to six riddles each with a different level of difficulty.

Each riddle was a one-paragraph description of a person. A sample is given below:

> Although my early teachers found me slow to learn, I was awarded the most famous award in my field just 20 years after the first time it was awarded! I have lived in several countries, such as Germany, Italy, Switzerland, and the USA. Out of my work, the most famous piece is also known as the world's most famous equation. I cannot say more about myself considering I was probably the most famous person of my line of work in the 20th Century.
>
> Who am I?

In the second part, workers were asked to count the number of cells visible in a given picture. Again six images with different levels of difficulties were used (Fig. 6.2). As the correct answer for each question is known, the validity of worker's responses are evaluated. The first part, solving a riddle, was considered as a heuristic task in which one needs to think and find a solution creatively. Based on the theory,

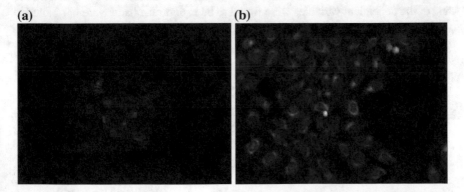

Fig. 6.2   Participants were asked to count the number of visible cells in the given pictures: **a** easy, **b** difficult

autonomously motivated participants should present better performance in such a task than controlled motivated ones (cf. Sect. 2.1.3). The second part, counting the number of cells, was considered as an algorithmic task in which a particular procedure needs to be followed. Based on the theory, autonomously motivated workers should perform as good as controlled motivated workers. Meanwhile, workers were asked to rate interestingness of each part and difficulty level of each riddle on a 5-point Likert scale. From the qualified group of workers, 149 individuals participated in the study; all had previously answered the CWMS questionnaire. The job was compensated with $2.

6.1.2.2 Results

Results of CWMS questionnaire showed that all participants were strongly driven by controlled motivation ($M = 5.08$, $SD = 0.9$). Overall 53% of the riddles and cell-counting questions were answered correctly. However, 85.2% of participants responded the cell-countering task with one or more errors and 79.9% of them had at least one error within their responses to the riddles. There was no correlation between motivation scores and the number of errors in the cell-counting task. However, a small but significant correlation between participant's intrinsic motivation score and number of errors they made in the riddle task was observed: $r(149) = .221$, $p = .007$. This contradicts with the hypothesis.

Investigating answers given to each question showed that perceived difficulty significantly correlates with the deviation of given answers from the ground truth in the cell-counting questions $r(893) = .408$, $p < .001$.

A Binomial Logistic Regression model was used to predict worker's performance (i.e. weather they answer accurately) given workers' underlying motivation scores, perceived interestingness, and difficulty of a question. The logistic regression model delivered an accuracy of 72.1% (sensitivity: 66.9%, specificity: 76.7%), and was found to be significant with $\chi^2(8) = 28.22$, $p < .001$ and explained 31.7% (Nagelkerke R^2) of variance. Besides the interestingness and difficulty of the question, the rest of the dependent variables were found to be insignificant and were not included (Table 6.2).

Table 6.2 Logistic regression model to predict error occurrence in cell counting and riddle solving task

Predictor	β	SE β	Wald's χ^2	df	p	e^β
Constant	−1.342	0.202	44.221	1	<.001	0.261
Interestingness	−.214	0.033	41.755	1	<.001	.807
Difficulty	.704	0.039	332.12	1	<.001	2.023

6.1.2.3 Discussion

Results of this study showed that perceived task attributes, namely interestingness and difficulty, can partly predict error occurrence in the responses, whereas underlying motivation of workers does not influence the accuracy of the outcome. Perceived interestingness is positively, and perceived difficulty level is negatively related to the accuracy of answers.

Meanwhile, results showed that workers with higher intrinsic motivation delivered less accurate results in the heuristic task. That contradicts the theory as the opposite was expected. However, results of this study should be interpreted with caution. As reported, from the invited list of participants, workers who were mostly driven by controlled motivation participated in the study. It could be due to the fact that the given task was not phrased as a meaningful and valuable occupation rather like an unwise activity which was highly paid out. Therefore, results can not be generalized to all workers.

6.1.3 Discussion

In this section the relation between the underlying crowdworking motivation of workers and their performance was investigated. Worker's participation and the reliability of their responses to the survey job were associated with the type and amount of motivation. However, between motivation subscales and the overall HIT approval rate no (or weak) correlations were observed. As this statistic is widely used by requesters to specify the group of workers who are allowed to work on their task, it is expected that all workers, including the externally motivated workers, keep their approval rate statistic high, to have access to more jobs.

Furthermore, results from Study 6.1 showed that the underlying crowdworking motivation of participants did not relate to the accuracy of their responses within the used algorithmic and heuristic task. Rather, perceived difficulty and interestingness of the task were associated with the accuracy of the replies. However, results can not be generalized for all workers as participants were highly driven by controlled motivation.

Follow-up studies should consider the interaction between worker's underlying crowdworking motivation and their task-specific motivation. For instance, how would an individual with intrinsic motivation, as the main underlying crowdworking motivation, perform when the given task is intrinsically (e.g. joyful), identified (e.g. believing in the value of activity) or externally (e.g. highly beneficial) motivating.

6.2   Task Design

In this section the effect of task design on the reliability of responses collected through crowdworking is investigated. The first study demonstrated that workers improve their performance due to the awareness that a reliability check method was integrated into the job. The effect was further analyzed in the second study to find out what attributes of a reliability check method can encourage workers to further improve their performance. This section is based on the previously published results in [82, 83].

6.2.1   Study 6.2: Effect of Being Observed on the Reliability of Responses

6.2.1.1   Method

The reliability of responses collected in two crowdsourcing survey studies was compared (Fig. 6.3). Both survey studies have been conducted with MTurk in order to develop the CWMS questionnaire. Beside others, they contained questionnaire items which were presented in the form of a 7-point Likert scale. Two reliability check methods were used, namely the Inconsistency Score (IS), and the Trapping Questions. In the IS method some randomly selected items are asked twice in the questionnaire. These items will be called consistency check items and are an exact duplicate of the original items which are positioned in a moderately far distance from the original ones. In general, the IS shows the overall response consistency of an individual participant and may be interpreted as an indicator of thoughtfully subjective response rather than randomly answering. The trapping questions method, is simply about employing statements with clear answers presented alike to the questionnaire items e.g. I believe two plus five does not equal nine. Such a question acts as a gold standard, and it is expected to indicate whether or not a worker consciously read the statements. The IS method was used in both surveys but the trapping questions method was used in the second survey only. A detailed description of both methods was given in the Sect. 4.2.1.1.

The first survey (hereafter "without trapping") contained 74 items; from them seven items were consistency check items for the IS method. Overall 256 crowdworkers responded to the survey, of which 222 answers were completely filled in. The second study (hereafter "with trapping"), contained 14 reliability check items including 6 consistency check items, for the IS method, and 8 trapping questions. The first trapping question was positioned as the third survey question to ensure that the workers recognize the presence of a reliability check method; the others were distributed across the survey. It is assumed that the trapping question method is easily noticeable by workers, whereas the presence of the IS method is almost unnoticeable for workers. Overall 405 crowdworkers responded to this survey from which 401 answers were complete. In both studies, the survey was only available for workers from the U.S with a minimum of 100 approved assignments. In addition, both surveys were published for different groups of workers, which were created based on their Approval Rate statistic (i.e. indicates worker's long-term reliability of responses). This was done to ensure that not only highly reliable workers participate in the study. Based on the approval rate statistics five groups were created (cf. Table 6.3).

6.2.1.2 Results

Figure 6.4 illustrates the distribution of the Inconsistency Scores for both surveys. The Kolmogorov-Smirnov test showed that the IS distributions are significantly difference from the normal distribution: $D_{NoTrapping}(222) = 0.96$, $p < .001$; $D_{WithTrapping}(401) = 0.13$, $p < .001$. A Mann-Whitney-U test showed that the distribution of Inconsistency Scores in the survey without trapping question ($Mdn = 1$) was significantly different from the distribution of Inconsistency Scores in the survey with trapping question ($Mdn = 0.65$), $U = 28025.5$, $z = -7.66$, $p < .001$, $r = -.31$.

Table 6.3 Number of participants in two surveys within Study 6.2

Study	Groups (based on approval rate)				
Without trapping	Range [%]	[0, 70]	(70, 80]	(80, 90]	(90, 95]	(95, 100]
	# workers	11	2	29	110	70
With trapping	Range [%]	[0, 85]	(85, 90]	(90, 95]	(95, 98]	(98, 100]
	# workers	58	80	92	93	78

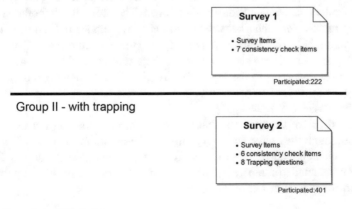

Group I - without trapping

Survey 1
• Survey Items
• 7 consistency check items

Participated:222

Group II - with trapping

Survey 2
• Survey Items
• 6 consistency check items
• 8 Trapping questions

Participated:401

Fig. 6.3 Procedure of Study 6.2

Fig. 6.4 Distribution of IS
in both surveys using all
responses, after [83]

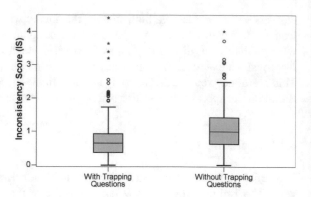

In addition, by re-arranging the approval rate groups, three groups with equivalent ranges between two surveys were created (i.e. [0, 90], (90, 95], and (95, 100]). In the survey without trapping 19%, 49.5%, and 31.5% of participants and in the survey with trapping 34.4%, 23%, and 42.6% of participants belong to the new groups, respectively. There was a weak negative correlation between the participants' approval rate (i.e. group) and the Inconsistency Score ($r_s = -.14$, $p < .001$) (Fig. 6.4).

6.2.1.3   Discussion

Two different measurements for evaluating reliability of responses in crowdsourcing surveys have been applied into two survey studies. The IS method has been used as an almost unnoticeable system for users whereas trapping questions were used in a way, which made the presence of quality control system obvious to the users. A lower Inconsistency Score was observed for the survey in which an obvious quality control system was used. One hypothesis is that recognizing the presence of a quality control system positively alters behavior of participants. Note that as the number of participants in each survey and proportion of them in each approval rating group were not equivalent, no firm conclusions can be drawn. Therefore, this hypothesis and the underlying reasons for such a behavior need to be investigated in future studies. One explanation might be that workers may recognize the importance of reliable answer for the job provider, or they see a danger of being rejected and losing the monetary reward. It may also refer to the Social Facilitation Effect known from the field of psychology: people tend to perform better on simple tasks and worse on complex tasks in the presence of others [35].

One drawback of this study is that the same set of items was not used in both surveys (although they have been very similar). In addition, the proportion of participants in the approval-rating groups were not equivalent in both surveys The next study was conducted to compare the influence of different types of trapping questions on the performance of crowdworkers.

6.2.2 Study 6.3: Effect of Trapping Questions on the Performance of Crowdworkers

In this study, the influence of trapping questions' design on the validity of the collected data through crowdsourcing was investigated. Crowdworkers were asked to assess the quality of speech samples from a standard database. In a between-group study design, each group was presented with different types of trapping questions, which were designed based on the results of Study 6.2 and state-of-the-art. The ratings obtained from the crowdworkers in each test group, were compared to ratings collected in the standard laboratory setting (assuming that laboratory ratings are equivalent or close to the ground truth).

6.2.2.1 Background

In the domain of telecommunication systems, the quality of transmitted speech as perceived by the user, the so-called Quality of Experience (QoE) [78], is assessed by service providers to optimize their services. Typically, listening-only-tests (LOTs) [110] methods are used for the QoE assessment of transmitted speech. In LOTs, naïve participants assess multiple stimuli which represent possible degradations of a transmission system in a standard laboratory environment. Typically, feedback is collected on a 5-point Absolute Category Rating (ACR) scale [110]; the average score for each stimulus over ratings of all test participants is called a Mean Opinion Score (MOS). Such lab-based LOTs provide reliable, valid results and are often used as the ground truth for research and industry [82]. A growing body of research addressed utilizing the crowdsourcing microtask for conducting QoE studies in various domains and pointed out differences [77, 80] and best practices [44–46]. Hoßfeld and Keimel [46] found that the correlations between the results from the laboratory and unpaid crowdworkers (i.e. recruited via social networks) are similar to the correlation between the lab studies. They conclude that appropriate mechanisms like reliability check and training phases should be included in the crowdsourcing task design.

Results from Study 6.2 showed that the presence of an obviously recognizable reliability check method in a crowdsourcing survey significantly increases the consistency of answers. It was hypothesized that workers may recognize the importance and value of a reliable answer for the employer and are thus more motivated to comply with the instructions and refrain from cheating [82]. A similar phenomenon was observed by Kittur et al. [62] when collecting workers ratings and feedbacks regarding the quality of Wikipedia articles. They concluded that the quality of the responses increases when the effort to conceal the act of cheating is as high as the effort to provide reliable answers. Meanwhile, Egelman et al. [25] pointed out that supervision and the face-to-face contact between experimenter and participants in the lab tests may encourage the participants to provide 'good' results.

Based on researches, as mentioned above, different types of trapping questions were created to test the following hypotheses:

- **H1**: The presence of trapping questions increases the quality of responses collected in a crowdsourcing environment using a LOT quality assessment task.
- **H2**: A trapping question is more effective when it makes participants aware of the importance of their work (based on Herzberg's two-factor theory of job satisfaction cf. Sect. 2.2).
- **H3**: A trapping question is more effective when the effort of concealing the cheating would be as high as the effort of providing reliable answers (based on [62]).

6.2.2.2   Method

Crowdworkers were randomly assigned to four experiment groups in which they assess quality of speech stimuli from a standard dataset. For each group different speech-related trapping questions were integrated into their task. The abovementioned hypotheses were evaluated by comparing the MOS ratings obtained in the crowdsourcing environment from each group with MOS values provided in the standard dataset collected in the lab setting. The MOS values collected in the lab setting are assumed to be equivalent or close to the ground truth. Therefore, such a comparison indicates the validity of the collected data in each study group (cf. Sect. 2.1.3). The study was conducted using the Crowdee mobile crowdsourcing platform [81]. Crowdee workers are mainly students from various places in Germany (cf. Chap. 3). Crowdee's in-built functionalities for media playback, automatic job-chains, temporal qualification, and randomization were used.

In the following, the employed standard dataset, the study procedure, the trapping questions and the data collection are described in detail.

Database (SwissQual 501)

For the experiment, we used 200 stimuli from the database number 501 from the ITU-T Rec. P.863 competition which has been provided by SwissQual AG, Solothurn. Overall 200 stimuli were used to carry 50 conditions. Each condition describes one degradation or a combination of degradations and each is composed of four stimuli (all with the same combination of degradations) recorded by four native German speakers with four different sentences. The conditions represent degradations like mixed audio bandwidth (narrowband 300–3400 Hz, wideband 50–7000 Hz, superwideband 50–14000 Hz), signal-correlated as well as uncorrelated noise, ambient background noise of different types (see [82] for more details). The database contains 24 quality assessments from German natives per stimulus, which were obtained in a standard laboratory environment according to ITU-T Rec. P.800. The resulting MOS per stimulus and test condition serves as a reference for this experiment.

Study procedure

Figure 6.5 illustrates the procedure of the experiment. Workers have performed three individual jobs one after each other: (1) Qualification, (2) Training, and (3)

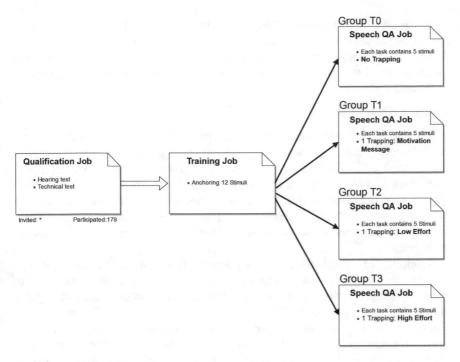

Fig. 6.5 Procedure of Study 6.3

Speech quality assessment. The qualification job was a short test including questions about hearing impairment and prior experiences with quality assessment, as well as a technical headset and audio playback test. Workers were asked to perform all tasks in a quiet place and to use two-eared headsets. It was forced such that all jobs could only be started when a headset is connected and validated by a simple math question in which digits swept between left and right in stereo. After approval, workers were assigned to one of the four different experimental groups (i.e. T0–T3) and automatically got access to the training job. There, crowdworkers performed a training job that contained twelve stimuli representing anchor types of degradations. Here workers were explicitly made aware of the presented degradation types. After successfully performing the training task, a temporary-expiring certificate (qualification) was issued with one-day validity duration. As a result, a worker was qualified to perform the speech quality assessment task as long as the certificate was valid. On its expiry, the worker was automatically returned to the training job for retraining. The speech quality assessment jobs were structured as follows: first workers were asked to report the level of background noise and their current level of tiredness. After that, depending on the study group they were assigned to, either five stimuli (no trapping group) or six stimuli (five stimuli plus one trapping stimulus) were presented to them to rate the quality of each of the non-trapping stimuli. For the trapping stimulus, the answer scales may differ depending on the study group. Thus, each job contained either five or six stimuli and workers were allowed to perform up to 40 jobs in a row (i.e. rate all 200 stimuli of the database), or pause in between tasks at their preference. Ratings were collected on a 5-point ACR scale similar to the scale used in the lab study. Note that, workers were forced to listen to the entire stimulus before they could proceed to the next one.

Trapping Questions

Three different study groups, each treated by a different configuration of speech-related trapping question, were compared to a control group with no trapping question (*Trapping T0—No Trapping*). For the study groups *T1–T3*, the speech quality assessment jobs were slightly altered by adding one additional stimulus, which was modified (i.e. trapping stimulus). From the original dataset, 40 different stimuli were randomly selected (different speakers, and different degradation conditions). Those stimuli were manipulated to create various types of audio trapping questions, for each group:

Trapping T1—motivation message. For the first group of trapping stimuli, a message was recorded with a speaker not being part of the speech material to be judged in German. It was appended to the first four seconds of each of the 40 trapping stimuli. The message was as follows: *"This is an interruption. We—the team of Crowdee—like to ensure that people work conscientiously and attentively on our tasks. Please select the answer X' to confirm your attention now."* In this group, the trapping question was visually identical to the other quality assessment questions, but the trapping stimulus asked workers to choose a specific answer option (e.g. $X = Poor$, or *Fair*).

Trapping T2—low effort. For this group, the 40 selected stimuli were manipulated by inserting an animal sound either in the middle or at the end of them. In this case, the trapping question was also visually different: Workers were asked to indicate in a multiple-choice question the particular animals that their noises were played within the current stimulus. Here, workers can provide a true answer for the trapping with low effort; full concentration in entire assessment job is not required.

Trapping T3—high effort. In the last group of trapping questions, the same stimuli created for the second group (*Trapping T2*) were employed, but the trapping stimulus was presented together with the ACR rating scale, which was used for all other stimuli. Additionally, at the end of the job (i.e. after being presented with five non-trapping stimuli plus one trapping stimulus) a multiple-choice question was added. It requested workers to specify all the animal sound(s) that they recognized in any of the previous (i.e. six) stimuli. Comparing to the *Trapping T2* group, where the trapping question was visually recognizable, here workers did not know where the animal sound appears unless they carefully listened to all stimuli. In case the worker was inattentive while rating the other stimuli, he/she would need to review all previous stimuli again, to find out the correct answer. In this case, the effort of concealing cheating is as high as performing the job correctly.

The objective to employ these kinds of trapping questions was to evaluate the abovementioned hypotheses: Comparing the *No Trapping T0* group with all the others showed whether the presence of trapping question has an effect (i.e. *H1*). The group

Trapping T1 emphasizes the importance and value of highly reliable responses to the workers (i.e. *H2*). With *Trapping T2* and *Trapping T3* we examined the assumption that the likelihood of cheating decreases if the effort to conceal cheating is as high as the effort to accurately complete a task (i.e. *H3*). For all groups, the position of the trapping stimuli in the task was shuffled each time. Moreover, it was planned to collect 24 ratings per stimuli within each study group similar to the laboratory dataset.

6.2.2.3 Results

The study was conducted using Crowdee for 18 days. After performing the qualification job, a worker was randomly assigned to one of the four experimental groups (*T0, T1, T2, T3*) by the platform. Thus, each worker could only perform the speech quality assessment jobs designed for their study group (between-group design). Overall, 179 workers (87 Female, $M_{age} = 27.9y., SD_{age} = 8.1y.$) participated in the study. Based on the trapping questions, 49 responses were rejected (*T1 = 2, T2 = 1, T3 = 46*). Note that, the responses from the task containing wrongly answered trapping questions were removed and other judgments from the same worker in tasks with correctly answered trapping questions were used.

For all experimental groups (*T0, T1, T2, T3*), MOS values were calculated for each of the 200 stimuli of the database. Four each group, a Spearman's rank-order correlation was computed to determine the relationship between the MOS values obtained from the crowd and the MOS values known from the laboratory experiment (i.e. provided within dataset). In addition, the Root Mean Square Deviations (RMSD) from the laboratory MOS values were calculated for the MOS values of the four experiment groups (Table 6.4 and Fig. 6.6).

Strong positive correlations with the MOS values obtained in the lab were observed for all experiment groups, regardless of the kind of employed trapping question. However, for *Trapping T1* the highest correlation, as well as the lowest RMSD, was observed; hence, this group shows the best performance.

Table 6.4 Correlation between the MOS ratings obtained in the lab and the MOS ratings obtained via crowdsourcing (N=200)

Group	r_s	p-value	RMSD
Trapping T0—No trapping	0.886	<0.001	0.426
Trapping T1—Motivation message	0.909	<0.001	0.375
Trapping T2—Low effort	0.897	<0.001	0.411
Trapping T3—High effort	0.909	<0.001	0.390

Table 6.5 Number of conditions with the CIs of the crowd means being lower, higher and overlapping with the CIs of the lab means

Group	N of CIs lower	N of CIs higher	N of CIs overlapping
Trapping T0—No trapping	17	6	27
Trapping T1—Motivation message	13	2	35
Trapping T2—Low effort	17	3	30
Trapping T3—High effort	16	4	30

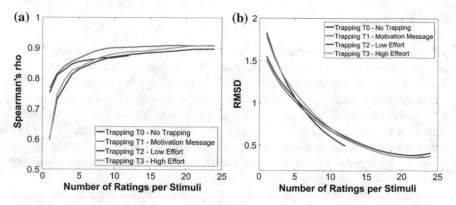

Fig. 6.6 Comparing MOS ratings obtained from crowdsourcing studies and MOS values from lab study, after [82]. Changes in correlation (**a**) and RMSD (**b**) depending on number of ratings

Next, for each of the 50 degradation conditions, MOS and its 95% confidence interval[2] (CIs) values obtained in each experiment group were calculated. Based on this data, the number of degradation conditions in which the CIs of the crowdsourcing ratings did not overlap with the CIs of the ratings obtained in the lab were counted. Again the results (cf. Table 6.5) showed best results for *Trapping T1* in which for 35 of the conditions an overlap of the CIs was observed. For both, *Trapping T2* and *Trapping T3*, for 30 conditions the CIs of the means were overlapping with CIs of the lab means. Poorest performance was observed for *Trapping T0*. Later, we examined if the number of overlapping and non-overlapping conditions for *T1, T2, and T3* differed from the control condition *T0*. A χ^2−test indicated a statistically significant difference between *T0* and *T1*, $\chi^2(1, N = 50) = 5.15$, $p = .023$. Accordingly, the results obtained with experiment group *Trapping T1*—Motivation Message are more consistent to the lab results than the results obtained with *Trapping T0*—*No Trapping*.

---

[2]The CIs can be used, like p-values, to estimate the statistical significance of an effect (e.g. non-overlapping 95% CIs indicate a difference on the $p < .01$ level) [16].

6.2.2.4 Discussion

The influence of different types of trapping questions on the reliability of speech quality crowdtesting is examined. In all groups with trapping questions (*T1, T2* and *T3*) all obtained data (correlations, RMSD, number of conditions for which the CIs of the means were overlapping with the CIs of the lab means) tended to be more consistent to the lab data compared to the data obtained in the group without any trapping question. As a result, the presence of trapping question increases the reliability of collected data through crowdsourcing. Best results were observed for the type of trapping question, for which a recorded voice was presented in the middle of a random stimulus. The voice explained to the workers that high quality responses are important, and asked them to select a specific item to show their concentration. A possible explanation for the effect of this kind of trapping questions is that they communicate the importance and the value of their work to the crowdworkers. Based on Herzberg's two-factor theory of job satisfaction, the presence of factors, such as acknowledgment and the feeling of being valued, facilitates satisfaction and autonomous motivation, which eventually leads to better performance.

6.3 Chapter Discussion and Summary

The effect of workers' underlying crowdworking motivation, tasks attributes, and tasks design (applying a reliability check method) on different measures of worker's performance was investigated.

A survey study showed that underlying autonomous crowdworking motivation is positively related to how active workers are on the platform. The consistency of responses observed in the same study was associated to how internalized their underlying motivation is. The long-term reliability of responses (indicated by HIT approval rate) was related to both identified and external regulatory motivation. That is feasible as the HIT approval rate is a statistic widely used by requesters to specify the workers who are qualified to work on their task. Therefore, it is expected that all workers who want to have access to more jobs, keep their approval rate statistic high. Note that, this statistic does not indicate how good workers performed their jobs, but it refers to the percentage of jobs whose response was good enough to not be rejected.

Next, the performance of participants was observed in algorithmic, and heuristic tasks. Results showed that perceived task attributes, namely interestingness and difficulty, can partly predict error occurrence in the responses whereas their underlying crowdworking motivation does not influence the accuracy of their response. As participants in the study were mostly driven by the controlled motivation, results can not be generalized to all workers.

Furthermore, the influence of employing reliability check methods on the performance of participants was investigated. The reliability check methods, are not only beneficial to find out inaccurate responses in the post-processing step, but also encourage workers to perform adequately. Comparing two survey studies showed that employing trapping questions in a noticeable way for workers leads to higher consistency in their responses. Results from the follow-up study also confirm that applying trapping questions, which encourage workers to perform more adequately, resulted in higher correlation with the laboratory data in a crowdtesting speech quality assessment task compared to a control group. Furthermore, a trapping question which communicates importance and value of work leads to the most equivalent results to the lab, followed by the trapping questions that the effort of concealing the cheating would be as high as the effort of providing reliable answers.

Consequently, employing reliability check methods is strongly recommended, especially in the form that workers partly notice them, and they communicate importance and value of their work to them. The studied task attributes, i.e. perceived interestingness and difficulty, influence the accuracy of answers collected in crowdsourcing. Therefore, it is suggested to communicate clearly and upfront to workers what qualifications are needed to perform the task. It is again important that workers understand that their activity is valuable apart from its direct monetary rewards.

Chapter 7
Conclusion and Outlook

7.1 Conclusion

This dissertation aims at answering questions about the motivation of crowdworkers, factors that are influencing it, and aspects affected by it. In Chap. 2, a taxonomy of crowdworking motivation was introduced based on the Self-Determination Theory of motivation. It distinguished between the general underlying motivation of crowdworking and a task-specific motivation which later refers to the motivation of choosing a particular task to perform instead of the other tasks. It was assumed that the general motivation influences the task-specific motivation. The taxonomy pointed out that user, task and platform characteristics may influence the motivation of workers. Also, the type and amount of motivation can affect workers' well-being and performance, namely their participation and reliability of their responses.

Chapter 3 aimed to understand who are the crowdworkers, specifically considering their demographic data. Empirically collected data showed that crowdworkers' population vary not only between platforms but also within platforms. Workers share some characteristics with the people of their home country like the household size but differ in many of them like education level, and age group, especially in developing countries. Other than that, the time spent for crowdworking relates to the crowdworkers' expenditure purpose of crowdworking income, and their household per capita income. It was also observed that results from a single survey varied in some cases from results of long-term repeated sampling of crowd population. This shows that proper sampling methods in each study should be considered to achieve a representative group of participants and be able to generalize findings.

Next, Chap. 4 focused on the general underlying motivation of crowdworkers and described the development process of the CWMS questionnaire. The questionnaire is a tailored instrument for measuring the general motivation of crowdworkers based on the Self-Determination Theory. Findings from the previous chapter were followed, and the questionnaire was developed using data from a representative sample of crowdworkers. Later, the CWMS was successfully validated using two other studies by participants from U.S and Indian population of MTurk crowdworkers.

© Springer International Publishing AG 2018
B. Naderi, *Motivation of Workers on Microtask Crowdsourcing Platforms*, T-Labs
Series in Telecommunication Services, https://doi.org/10.1007/978-3-319-72700-4_7

In Chap. 5, task selection strategies of workers were studied in detail. The main assumption was that accepting to perform a task is an indicator of being motivated, i.e. presence of task-specific motivation. Five studies were conducted in MTurk. First, a quantitative study which found out that payment, requester's repetition, and task-specific factors including difficulty, complexity, and interestingness of a task are major factors for crowdworkers when deciding about accepting a HIT. The next study revealed that the RSME scale is an appropriate measure of estimated workload, and 20-points bipolar scales can be used for measuring other factors like interestingness. Later, in a large-scale study, a dataset with 401 HITs from different types was created in which each HIT was rated by 15 different crowdworkers. Results showed that the proposed logistic regression model could predict individual decisions on accepting an arbitrary HIT with an accuracy of 89.54%. The model is based on individual assessments of task's interestingness, expected workload, profitability (ratio of rewards to expected workload), and frequency of performing similar HITs. However, it was shown that using mean scores of measured variables drops the accuracy of prediction to 69.9%. This indicates that individual differences in skills and personal experiences, in addition to personal preferences, are influential factors which are not considered in the model. Meanwhile, it was observed that the importance of each factor changed depending on the worker's general underlying motivation of crowdworking, but the difference was not significant. Later, using an Adaptive Choice-Based Conjoint analysis, the relative importance of the task-related factor, requester's repetition, and payment were evaluated for U.S and Indian population of MTurk. Results showed that payment is the major factor for both populations on taking a task. From task-related factors, expected workload and interestingness are ranked next for U.S workers, but Indian workers valued the number of available HITs (the more, the better) and creation time of the HIT (the newer the HIT, the better) are as important as its interestingness. From requesters attributes, fairness and generosity were considered to be the most important aspects for both populations. Last but not least, a model was created for predicting the mean expected workload score of a HIT given its meta-data, and design (HTML code, and a screenshot). In 72% of times, the predicted value was in the range of 95% confidence intervals. Applying the predicted score for expected workload (and consequently profitability) into the previously introduced acceptance model leads to an accuracy of 77.21% on predicting individuals' decisions.

Moreover, Chap. 6 reported the effect of motivation and task design on the performance of crowdworkers. In the first study, results showed that the amount of workers' participation (self-reported) and reliability of their answers is positively related to their autonomous motivation scores. Identified and external motivation of workers significantly correlate with quality of their work in long-term. However, a follow-up study in which workers were asked to perform different tasks after filling the CWMS questionnaire, could not confirm the relation between the underlying motivation of crowdworkers and the reliability of their work. Instead, the task-specific factors, namely interestingness and difficulty of task, were the important factors which predict the reliability of answers. However, results of this study should be interpreted with caution as participants were not a representative group of workers considering their underlying motivation. Furthermore, two studies showed that using a reliability check method not only facilitates the data screening process but also it encourages workers to provide more reliable answers when it is acknowledging their works and emphasizes the importance of high-quality work.

7.2 Future Work

Primary, findings from this thesis should be evaluated on different platforms and with diverse demographics of workers. As shown in Chap. 3, crowdworkers' population vary not only between platforms but also within platforms. All the studies presented in Chaps. 4–6 were conducted on MTurk with the exception of Study 6.3. Therefore, results of this thesis need to be validated with other platforms to be able to generalize the results. In the following, an outlook for future research in three areas is given.

7.2.1 General Motivation

The research considering the general underlying motivation of crowdworkers can be extended to answer two important questions. First, further research is needed to investigate the influence of different combinations of the general motivation and task-specific motivation on the performance of workers. The question is, how does the general motivation of a worker influence his performance in a task that is mostly motivated by interestingness (intrinsic motivation), profitability (external motivation) or frequency (identified motivation) with different degrees of expected workload.

Proposition 7.1 *Workers with a specific major general motivation type perform better when the task is motivated by a same nature of motivation.*

Second, the influence of platform design on the general motivation of worker, and worker's subjective well-being need to be studied. Although both aspects were presented in the taxonomy in Sect. 2.1, they were not investigated in this dissertation. Based on the theory, when a platform is designed to be autonomy supportive, then workers further internalize the regulation of their activity. Autonomy supportive means that procedures, workflows, and the environment support satisfying the needs for autonomy, competence, and relatedness. Further research is needed to find out how to enhance crowdworking procedures to be Autonomy supportive.

Proposition 7.2 *Platform's design that facilitates satisfying of needs for autonomy, competence, and relatedness, leads to higher degree of identified general motivation, which corresponds to more participation, and higher reliability of outcomes.*

7.2.2 Task-Specific Motivation

The acceptance model can be extended to create: (1) an optimal task-scheduler for crowdworkers, (2) predict reasonable rewards given the job design which can be used by job providers or platforms, (3) monitor the crowdworking market for a fair working condition to be used by labor unions.

The following steps needs to be conducted to extend the acceptance model. First, the resulting model, which predicts the expected workload, should be extended by advanced semantic and visual features. Also, rewards should be removed from the list of independent factors used by the model. This model can be used for goal 2 and 3, which predict the expected workload for an average crowdworker. Similarly, two sub-models for predicting the interestingness and the task type given the task design and meta-data should be created. The values predicted by the sub-model for expected workload (and consequently profitability), interestingness, and frequency (through task type) should replace the individual ratings in the acceptance model.

Furthermore, the acceptance model needs to be improved in a way that prediction accuracy increases when mean (or predicted) values are used instead of the individual users' ratings. Thus, user characteristics like skills, experiences, and preferences can be added. Methods to measure those characteristics of a worker should be invented. As a result, the individual decision of each worker to perform a given task can be predicted.

Another aspect that needs further research is the interaction between estimated workload and perceived one. Given the task meta-data and the preview, screen workers estimate workload of the task. However, it could be a case that when performing the task, their perceived workload substantially differs from their expectation. What would be the consequences of such an experience?

Proposition 7.3 *When the perceived workload substantially differs from estimated workload of the task, a crowdwoker may withdraw or perform the work with less accuracy depending to his/her estimation of workload for the remainings part of activity.*

Of further interest is the possibility to predict such a difference or eventually prevent it. As a result, one category of ill-designed jobs can be automatically detected and prevented in the early stages. Note that currently some platforms use a group of senior crowdworkers to evaluate the design of any submitted job before publishing them publicly. Although such an approach appropriately works for crowdsourcing platforms, it can be partly supported by the above-mentioned model to reduce costs.

7.2.3 Performance

Performance, and especially the quality of the created data through crowdsourcing microtasks is and will be the major research question in the future of crowdworking. Methods aiming to enhance performance through workers' motivation should dynamically adapt themselves based on the status quo. An innovative motivation-enhancing method may lose its functionalities during the time by workers getting used to it. It can even lead to an opposite effect.

Proposition 7.4 *When a noticible reliability check method is used extensively by job providers, not only its positive influence on encouraging crowdwoker to give high-quality response will vanish, but also it makes them feeling controlled and consequently demotivates them.*

Although needs are universal, motives which satisfy those needs may lose their effect during the time. Therefore, looking for the needs is crucial. This direction of future research seems to be promising for a more complete understanding of *what motivates crowdworkers to provide high-quality responses.*

Appendix A
Features and the Model for Predicting Expected Workload

A.1 Features

In the following complete list of features extracted and used for building the expect workload prediction model (cf. Sect. 5.5) is given.

A.1.1 Syntax Based Features

Feature	Description
characters	Number of characters visible to user by first time loading the HTML
fileSize	Size of HTML file
a_count_external	Number of external <a> tags
a_count_internal	Number of internal <a> tags i.e. links to other elements in the same page
button	Number of <button> tags
div	Number of <div> tags
span	Number of <span> tags
p	Number of <p> tags
header_h1	Number of <h1> tags
header_h2	Number of <h2> tags
header_h3	Number of <h3> tags
header_h4	Number of <h4> tags
selection	Number of <selection> tags
input_count	Number of <input> tags, considering unique "name"
input_type_hidden	Number of <input type="hidden"> tags (not visible to user)

© Springer International Publishing AG 2018
B. Naderi, *Motivation of Workers on Microtask Crowdsourcing Platforms*, T-Labs
Series in Telecommunication Services, https://doi.org/10.1007/978-3-319-72700-4

input_type_button	Number of <input type="button"> tags
input_type_number	Number of <input type="number"> tags
input_type_checkbox	Number of <input type="checkbox"> tags, aggregated by "name"
input_type_radio	Number of <input type="radio"> tags, aggregated by "name"
input_type_submit	Number of <input type="submit"> tags
input_type_text	Number of <input type="text"> tags
input_type_url	Number of <input type="url"> tags
textarea	Number of <textarea> tags
sum_inputs	sum of all inputs which required user's action
img_count	Number of <img> tags
img_under100px	Number of <img> tags with width smaller than 100 pixel
audio	Number of <audio> tags
video	Number of <video> tags

A.1.2 Semantic Based Features

Feature	Description
sentences	Number of sentences visible to the user
subclauses	Number of subclauses visible to the user
words	Number of words in the text visible to the user
avgWordLength	Average word lenght
numUniqueStems	Number of unique stems
avgUniqueStems	Ratio of unique stems to the number of words

A.1.3 Visual Features

All of visual features were calculated based on an intermediate feature *Text density* which refers to ratio of the white pixels (text in the images) to the size of the window.

Feature	Description
threshold	Mean (over all windows) text density in the image
quartiles 25	25% of windows have less text density than or equal to this value
quartiles 50	50% of windows have less text density than or equal to this value

quartiles 75	75% of windows have less text density than or equal to this value
quartiles 95	95% of windows have less text density than or equal to this value
quartiles 98	98% of windows have less text density than or equal to this value
quartiles 100	100% of windows have less text density than or equal to this value
count_dense_windows	The number of dense windows
max_conn_region_w	Number of windows in the largest connected region
max_conn_region_p	Number of pixels in the largest connected region
avg_conn_region_w	The average size of connected regions (windows)
quar_conn_region 75	75% of connected regions contains less less windows than or equal to this value
quar_conn_region 98	98% of connected regions contains less less windows than or equal to this value

A.2 Prediction Model

The ensemble model, created for predicting the expected workload given the HIT design, consists of ten component models each with different predictors and corresponding coefficient. The result of the ensemble model is calculated by getting mean over results of all ten component models.

In the following the over all relative importance of top ten predictors, and coefficients of predictors on each component model are presented.

Predictor	Relative importance
HIT type	0.2489
quartiles 100	0.0685
quartiles 98	0.0595
quartiles 75	0.0554
quartiles 95	0.0537
Subclauses	0.0529
Reward	0.0452
numUniqueStems	0.0406
Words	0.0299
Threshold	0.0261

Predictor	Model 1	Model 2	Model 3	Model 4	Model 5	Model 6	Model 7	Model 8	Model 9	Model 10
Constant	2.195	1.744	1.545	1.747	1.984	0.512	2.329	1.895	2.107	0.409
[HIT Type] (SV)	−0.283	−0.374	−0.351	−0.346	−0.338	−0.100	−0.255	−0.263	−0.376	−0.218
[HIT Type] (CC, IF)	0.356	0.338	0.341	0.303	0.382	0.384	0.399	0.328	0.349	0.434
[HIT Type] (CA, IA, VV)	0.000	0.000	0.000	0.000	0.000	0.000	0.000	0.000	0.000	0.000
reward	1.266	1.217	1.536	1.547	1.330	1.405	1.407	1.251	1.458	1.557
Syntax based features										
a_count_external								-0.004		
button	−0.060		−0.041	−0.066	−0.103	−0.089	−0.080	−0.092	−0.052	−0.085
span					0.006					0.005
p			0.007		0.005			0.007		
div		−0.002		−0.003	−0.002		−0.003			
header_h1	−0.231			−0.122					−0.175	
header_h2		−0.134				−0.109		0.111		
header_h3		0.081	0.073	0.102	0.083	0.068		0.098	0.058	
header_h4		−0.066		−0.033	−0.053			0.035	−0.047	0.049
input_count	−0.006				−0.008					−0.004
[input_type_file=0]	−1.580	−1.355	−1.224	−1.462	−1.404		−1.664	−1.424	−1.482	
[input_type_file=2]	0.000	0.000	0.000	0.000	0.000		0.000	0.000	0.000	
[input_type_submit=0]	0.153	0.132	0.268	0.276	0.247	0.243	0.369	0.216	0.233	
[input_type_submit=1]	0.000	0.000	0.000	0.000	0.000	0.000	0.000	0.000	0.000	
input_type_button	0.606				0.283			0.694		0.335
input_type_checkbox				0.022	0.034			0.040		0.026
input_type_hidden			−0.009	−0.016			−0.015	−0.023	−0.010	
input_type_number						−0.290	−0.138		−0.167	
input_type_radio				0.052	−0.046					
input_type_text		0.007	0.028	0.017	0.020			0.018		0.012
input_type_url	−0.384	−0.529		−0.531	−0.249	−0.597	−0.311	−0.256	−0.499	−0.525
textarea_count					0.045	0.044	0.050	0.038		
selection							0.024	0.023		
sum_inputs			−0.005					−0.007		
img_count								−0.013		
img_under100px	0.050	0.097		0.073			0.100		0.061	
video	0.335	0.343	0.277	0.453	0.399	0.351	0.530	0.447	0.723	0.320
Semantic based features										
avgUniqueStems	2.257		1.238			2.452	2.285		2.089	1.745
avgWordLength		0.077	0.084	0.103						
numUniqueStems		0.004		0.006	0.004			0.007		
sentences							−0.002		−0.002	−0.002
Subclauses	0.009			0.003	0.004	0.010			0.004	0.006
words				0.000						
Visual features										
quartiles 25		12.996	14.922			8.736	13.050			
quartiles 50			−2.499							
quartiles 75			0.906		−0.927		−1.240	−1.057		0.775
quartiles 95					0.987					
quartiles 98				0.336						
quartiles 100	1.031	0.983				0.674	0.480	0.744	0.489	0.905
max_conn_region_w									−0.285	
avg_conn_region_w			0.083			0.075		0.086	0.349	
threshold	−2.275	−1.646	−1.181			−2.046			−1.325	−1.450

References

1. Amitabh Mall Rachit Mathur, N.P.: Decoding digital @ retail: Winning the omnichannel consumer with digital in retail (2017)
2. Antin, J., Shaw, A.: Social desirability bias and self-reports of motivation: a study of Amazon mechanical turk in the U.S and India. In: Proceedings of the SIGCHI Conference on Human Factors in Computing Systems, pp. 2925–2934, CHI 2012 (2012). https://doi.org/10.1145/2207676.2208699
3. Baard, P.P., Deci, E.L., Ryan, R.M.: Intrinsic need satisfaction: a motivational basis of performance and weil-being in two work settings1. J. Appl. Soc. Psychol. **34**(10), 2045–2068 (2004)
4. Bederson, B.B., Quinn, A.J.: Web workers unite! addressing challenges of online laborers. In: CHI 2011 Extended Abstracts on Human Factors in Computing Systems, pp. 97–106. ACM (2011)
5. Berg, J.: Income security in the on-demand economy: findings and policy lessons from a survey of crowdworkers. Comparative Labor Law Policy J. **37**(3) (2016)
6. Borst, I.: Understanding crowdsourcing: effects of motivation and rewards on participation and performance in voluntary online activities. EPS-2010-221-LIS (2010)
7. Brawley, A.M., Pury, C.L.: Work experiences on MTurk: job satisfaction, turnover, and information sharing. Comput. Hum. Behav. **54**, 531–546 (2016)
8. Breiman, L.: Bagging predictors. Machine learning **24**(2), 123–140 (1996)
9. Buhrmester, M., Kwang, T., Gosling, S.D.: Amazon's mechanical turk a new source of inexpensive, yet high-quality, data? Perspectives Psychological Sci. **6**(1), 3–5 (2011)
10. Byrne, B.M.: Structural Equation Modeling with Amos Basic Concepts, Applications, and Programming. Routledge (2010)
11. Casner, S.M., Gore, B.F.: Measuring and evaluating workload: a primer. NASA Technical Memorandum (2010-216395) (2010)
12. Chandler, D., Kapelner, A.: Breaking monotony with meaning: motivation in crowdsourcing markets. J. Economic Behav. Organization **90**, 123–133 (2013)
13. Chandler, J., Mueller, P., Paolacci, G.: Nonnaïveté among amazon mechanical turk workers: consequences and solutions for behavioral researchers. Behav. Res. Methods **46**(1), 112–130 (2014)
14. Chapman, C.N., Alford, J.L., Johnson, C., Weidemann, R., Lahav, M.: CBC versus ACBC: comparing results with real product selection. In: 2009 Sawtooth Software Conference Proceedings, pp. 25–27 (2009)
15. Chen, J.: Enhanced recursive feature elimination. In: Machine Learning and Applications, ICMLA 2007 (2007)
16. Cumming, G., Finch, S.: Inference by eye: confidence intervals and how to read pictures of data. Am. Psychol. **60**(2), 170 (2005)

© Springer International Publishing AG 2018
B. Naderi, *Motivation of Workers on Microtask Crowdsourcing Platforms*, T-Labs Series in Telecommunication Services, https://doi.org/10.1007/978-3-319-72700-4

17. Dai, P., Rzeszotarski, J.M., Paritosh, P., Chi, E.H.: And now for something completely different: improving crowdsourcing workflows with micro-diversions. In: Proceedings of the 18th ACM Conference on Computer Supported Cooperative Work & Social Computing, pp. 628–638. ACM (2015)

18. De Waard, D.: The Measurement of Drivers' Mental Workload. Groningen University Traffic Research Center, Netherlands (1996)

19. Deci, E.L., Ryan, R.M.: Intrinsic Motivation and Self-determination in Human Behavior. Plenum (1985)

20. Deci, E.L., Ryan, R.M.: The "what" and "why" of goal pursuits: human needs and the self-determination of behavior. Psychological Inquiry 11(4), 227–268 (2000). https://doi.org/10.1207/S15327965PLI1104_01

21. Difallah, D.E., Catasta, M., Demartini, G., Ipeirotis, P.G., Cudr-Mauroux, P.: The dynamics of micro-task crowdsourcing: the case of Amazon MTurk. In: Proceedings of the 24th International Conference on World Wide Web, WWW 2015, pp. 238–247. ACM (2015). https://doi.org/10.1145/2736277.2741685

22. Doan, A., Ramakrishnan, R., Halevy, A.Y.: Crowdsourcing systems on the world-wide web. Commun. ACM 54(4), 86–96 (2011)

23. Doyle, C.: Work and Organizational Psychology: An Introduction with Attitude. Psychology at Work. Taylor & Francis (2004)

24. Efroymson, M.: Multiple regression analysis. Math. Methods Digital Comput. 1, 191–203 (1960)

25. Egelman, S., Chi, E.H., Dow, S.: Crowdsourcing in HCI research. In: Ways of Knowing in HCI, pp. 267–289. Springer, New York (2014)

26. Estellés-Arolas, E., González-Ladrón-De-Guevara, F.: Towards an integrated crowdsourcing definition. J. Inf. Sci. 38(2), 189–200 (2012)

27. Felstiner, A.: Working the crowd: employment and labor law in the crowdsourcing industry. Berkeley J. Employ. Labor Law, pp. 143–203 (2011)

28. Field, A.: Discovering Statistics using IBM SPSS Statistics, 4 edn. MobileStudy, SAGE (2013)

29. Finnerty, A., Kucherbaev, P., Tranquillini, S., Convertino, G.: Keep it simple: reward and task design in crowdsourcing. In: Proceedings of the Biannual Conference of the Italian Chapter of SIGCHI, p. 14. ACM (2013)

30. Gadiraju, U., Kawase, R., Dietze, S.: A taxonomy of microtasks on the web. In: Proceedings of the 25th ACM Conference on Hypertext and Social Media, HT 2014, pp. 218–223. ACM, New York, NY, USA (2014). https://doi.org/10.1145/2631775.2631819

31. Gadiraju, U., Kawase, R., Dietze, S., Demartini, G.: Understanding malicious behavior in crowdsourcing platforms: the case of online surveys. In: Proceedings of CHI, vol. 15 (2015)

32. Gagn, M., Deci, E.L.: Self-determination theory and work motivation. J. Organizational Behav. 26(4), 331–362 (2005). https://doi.org/10.1002/job.322

33. Gagn, M., Forest, J., Vansteenkiste, M., Crevier-Braud, L., Van den Broeck, A., Aspeli, A.K., Bellerose, J., Benabou, C., Chemolli, E., Gntert, S.T.: others: The multidimensional work motivation scale: validation evidence in seven languages and nine countries. Eur. J. Work Organizational Psychol. 24(2), 178–196 (2015)

34. Geiger, D., Seedorf, S., Schulze, T., Nickerson, R.C., Schader, M.: Managing the crowd: towards a taxonomy of crowdsourcing processes. In: AMCIS (2011)

35. Guerin, B.: Social facilitation. In: The Corsini Encyclopedia of Psychology. Wiley, New York (2010)

36. Hart, S.G.: NASA-task load index (NASA-TLX); 20 years later. In: Proceedings of The Human Factors and Ergonomics Society Annual Meeting, vol. 50, pp. 904–908. Sage Publications Sage, Los Angeles (2006)

37. Hart, S.G., Staveland, L.E.: Development of NASA-TLX (task load index): Results of empirical and theoretical research. Adv. Psychology 52, 139–183 (1988)

38. Hauser, J., Ding, M., Gaskin, S.P.: Non-compensatory (and compensatory) models of consideration-set decisions. In: 2009 Sawtooth Software Conference Proceedings, Sequin WA (2009)

39. Herzberg, F.: One more time: how do you motivate employees? Harvard Bus. Rev. **46**(1), 53–62 (1968)
40. Hill, S.G., Iavecchia, H.P., Bittner Jr., A.C., Byers, J.C., Zaklad, A.L., Christ, R.E.: Comparison of four subjective workload rating scales. Hum. Factors **34**(4), 429–439 (1992)
41. Hirth, M., Hoßfeld, T., Tran-Gia, P.: Anatomy of a crowdsourcing platform-using the example of microworkers.com. In: 2011 Fifth International Conference on Innovative Mobile and Internet Services in Ubiquitous Computing (IMIS), pp. 322–329. IEEE (2011)
42. Homburg, C., Giering, A.: Konzeptualisierung und operationalisierung komplexer konstrukte. ein leitfaden für die marketingforschung. Marketing-Zeitschrift für Forschung und. Praxis **18**(1), 5–24 (1996)
43. Hoßfeld, T., Heegaard, P.E., Varela, M., Möller, S.: QoE beyond the MOS: an in-depth look at QoE via better metrics and their relation to MOS. Quality User Experience **1**(1), 2 (2016)
44. Hoßfeld, T., Hirth, M., Korshunov, P., Hanhart, P., Gardlo, B., Keimel, C., Timmerer, C.: Survey of web-based crowdsourcing frameworks for subjective quality assessment. In: 2014 IEEE 16th International Workshop on Multimedia Signal Processing (MMSP), pp. 1–6. IEEE (2014)
45. Hoßfeld, T., Hirth, M., Redi, J., Mazza, F., Korshunov, P., Naderi, B., Seufert, M., Gardlo, B., Egger, S., Keimel, C.: Best practices and recommendations for crowdsourced QoE - lessons learned from the qualinet task force crowdsourcing (2014)
46. Hoßfeld, T., Keimel, C.: Crowdsourcing in QoE evaluation. In: Quality of Experience, pp. 315–327. Springer, Cham (2014)
47. Hoßfeld, T., Seufert, M., Hirth, M., Zinner, T., Tran-Gia, P., Schatz, R.: Quantification of YouTube QoE via crowdsourcing. InMultimedia (ISM), 2011 IEEE International Symposium pp. 494–499 (2011)
48. House, R.J., Wigdor, L.A.: Herzberg's dual-factor theory of job satisfaction and motivation: a review of the evidence and a criticism. Pers. Psychol. **20**(4), 369–390 (1967)
49. Howe, J.: The rise of crowdsourcing. Wired Magazine **14**(6), 1–4 (2006)
50. Hu, L.t., Bentler, P.M.: Cutoff criteria for fit indexes in covariance structure analysis: conventional criteria versus new alternatives. Structural Equ. Modeling Multidisciplinary J. **6**(1), 1–55 (1999)
51. Hurvich, C.M., Tsai, C.L.: Regression and time series model selection in small samples. Biometrika, pp. 297–307 (1989)
52. Iacobucci, D.: Structural equations modeling: fit indices, sample size, and advanced topics. J. Consumer Psychology **20**(1), 90–98 (2010). https://doi.org/10.1016/j.jcps.2009.09.003
53. Ipeirotis, P.: Demographics of mechanical turk: now live! (April 2015 edition) (2015). http://www.behind-the-enemy-lines.com/2015/04/demographics-of-mechanical-turk-now.html
54. Ipeirotis, P.G.: Demographics of mechanical turk (2010)
55. Ipeirotis, P.G.: Analyzing the Amazon mechanical turk marketplace. XRDS **17**(2), 16–21 (2010-12). https://doi.org/10.1145/1869086.1869094
56. Irani, L.C., Silberman, M.S.: Turkopticon: interrupting worker invisibility in Amazon mechanical turk. In: Proceedings of CHI 2013, pp. 611–620. ACM (2013). https://doi.org/10.1145/2470654.2470742
57. Kahneman, D., Diener, E., Schwarz, N.: Well-Being: foundations of hedonic psychology. Russell Sage Foundation (1999)
58. Kaufmann, N., Schulze, T., Veit, D.: More than fun and money. Worker motivation in crowdsourcing - a study on mechanical turk. AMCIS (2011)
59. Kazai, G., Kamps, J., Milic-Frayling, N.: The face of quality in crowdsourcing relevance labels: demographics, personality and labeling accuracy. In: Proceedings of the 21st ACM International Conference on Information and Knowledge Management, pp. 2583–2586. ACM (2012)
60. Keimel, C., Habigt, J., Horch, C., Diepold, K.: Qualitycrowd–a framework for crowd-based quality evaluation. In: Picture Coding Symposium (PCS), pp. 245–248. IEEE (2012)
61. Kinnaird, P., Dabbish, L., Kiesler, S., Faste, H.: Co-worker transparency in a microtask marketplace. In: Proceedings of the 2013 Conference on Computer Supported Cooperative Work, pp. 1285–1290. ACM (2013)

62. Kittur, A., Chi, E.H., Suh, B.: Crowdsourcing user studies with mechanical turk. In: Proceedings of the SIGCHI Conference on Human Factors in Computing Systems, CHI 2008, pp. 453–456. ACM (2008). https://doi.org/10.1145/1357054.1357127

63. Kittur, A., Nickerson, J.V., Bernstein, M., Gerber, E., Shaw, A., Zimmerman, J., Lease, M., Horton, J.: The future of crowd work. In: Proceedings of the 2013 Conference on Computer Supported Cooperative Work, p. 1301. ACM Press (2013). https://doi.org/10.1145/2441776.2441923

64. Koestner, R., Losier, G.F., Vallerand, R.J., Carducci, D.: Identified and introjected forms of political internalization: extending self-determination theory. J. Pers. Soc. Psychol. **70**(5), 1025 (1996)

65. Lam, C.F., Gurland, S.T.: Self-determined work motivation predicts job outcomes, but what predicts self-determined work motivation? J. Res. Pers. **42**(4), 1109–1115 (2008)

66. Landis, J.R., Koch, G.G.: The measurement of observer agreement for categorical data. Biometrics, pp. 159–174 (1977)

67. Le Moan, S., Mansouri, A., Hardeberg, J.Y., Voisin, Y.: Saliency for spectral image analysis. IEEE J. Selected Topics Appl. Earth Observations Remote Sensing **6**(6), 2472–2479 (2013)

68. Loomis, J.B.: Strategies for overcoming hypothetical BIAS in stated preference surveys. J. Agric. Resour. Economics **39**(1), 34–46 (2014)

69. Lowry, P.B., Gaskin, J.: Partial least squares (PLS) structural equation modeling (SEM) for building and testing behavioral causal theory: when to choose it and how to use it. IEEE Trans. Professional Commun. **57**(2), 123–146 (2014)

70. Malone, T.W., Laubacher, R., Dellarocas, C.: Harnessing crowds: mapping the genome of collective intelligence. Massachusetts Institute of Technology (2009)

71. Manning, C.D., Surdeanu, M., Bauer, J., Finkel, J., Bethard, S.J., McClosky, D.: The Stanford CoreNLP natural language processing toolkit. In: Association for Computational Linguistics (ACL) System Demonstrations, pp. 55–60 (2014). http://www.aclweb.org/anthology/P/P14/P14-5010

72. Martin, D., Carpendale, S., Gupta, N., Hoßfeld, T., Naderi, B., Redi, J., Siahaan, E., Wechsung, I.: Understanding the crowd: ethical and practical matters in the academic use of crowdsourcing, pp. 27–69. Springer International Publishing (2017). https://doi.org/10.1007/978-3-319-66435-4_3

73. Martin, D., Hanrahan, B.V., O'Neill, J., Gupta, N.: Being a turker. In: Proceedings of the 17th ACM Conference on Computer Supported Cooperative Work and Social Computing, pp. 224–235. ACM (2014)

74. Mason, W., Watts, D.J.: Financial incentives and the performance of crowds. ACM SigKDD Explorations Newslett. **11**(2), 100–108 (2010)

75. McAuley, E., Duncan, T., Tammen, V.V.: Psychometric properties of the intrinsic motivation inventory in a competitive sport setting: a confirmatory factor analysis. Res. Q. Exercise Sport **60**(1), 48–58 (1989). https://doi.org/10.1080/02701367.1989.10607413

76. Miller, S.: Workload measures. National Advanced Driving Simulator. Iowa City, United States (2001)

77. Möller, S., Hoßfeld, T., Naderi, B.: Comments on P.CROWD (2016)

78. Möller, S., Raake, A.: Quality of Experience: Advanced Concepts. Applications and Methods. T-Labs Series in Telecommunication Services. Springer, New York (2014)

79. Murphy, J.J., Allen, P.G., Stevens, T.H., Weatherhead, D.: A meta-analysis of hypothetical bias in stated preference valuation. Environ. Resource Econ. **30**(3), 313–325 (2005)

80. Naderi, B., Möller, S.: Speech quality assessment in crowdsourcing: comparison with laboratory and recommendations on task design. Technical Report COM 12-C 039, Geneva, Switzerland (2017)

81. Naderi, B., Polzehl, T., Beyer, A., Pilz, t., Möller, S.: Crowdee: mobile crowdsourcing microtask platform—for celebrating the diversity of languages. In: Proceedings of 15th Annual Conference of the International Speech Communication Association (Interspeech 2014). IEEE (2014)

82. Naderi, B., Polzehl, T., Wechsung, I., Kster, F., Möller, S.: Effect of trapping questions on the reliability of speech quality judgments in a crowdsourcing paradigm. In: 16th Annual Conference of the International Speech Communication Association (Interspeech 2015), ISCA, pp. 2799–2803 (2015)

83. Naderi, B., Wechsung, I., Möller, S.: Effect of being observed on the reliability of responses in crowdsourcing micro-task platforms. In: 2015 Seventh International Workshop on Quality of Multimedia Experience (QoMEX), pp. 1–2. IEEE (2015). https://doi.org/10.1109/QoMEX. 2015.7148091

84. Naderi, B., Wechsung, I., Möller, S.: Crowdsourcing work motivation scale: development and validation for crowdsourcing micro-task platforms. In: Prepration (2016)

85. Naderi, B., Wechsung, I., Polzehl, T., Möller, S.: Development and validation of extrinsic motivation scale for crowdsourcing micro-task platforms. In: Proceedings of CrowdMM 2014, pp. 31–36. ACM (2014). https://doi.org/10.1145/2660114.2660122

86. Novotorova, N.: A Conjoint Analysis of Consumer Preferences for Product Attributes: The Case of Illinois Apples. University of Illinois at Urbana-Champaign (2007)

87. O'Neill, J., Martin, D.: Relationship-based business process crowdsourcing? In: Kotzé, P., Marsden, G., Lindgaard, G., Wesson, J., Winckler, M. (eds.) Human-Computer Interaction-INTERACT 2013, pp. 429–446. Springer, Heidelberg (2013)

88. Orme, B.K.: The cbc/hb system for hierarchical bayes estimation version 5.0 technical paper. Technical Paper Series, Sawtooth Software, Orem, UT (2009)

89. Orme, B.K.: Getting started with conjoint analysis: strategies for product design and pricing research. Research Publishers (2010)

90. Peer, E., Samat, S., Brandimarte, L., Acquisti, A.: Beyond the turk: an empirical comparison of alternative platforms for crowdsourcing online behavioral research **2594183** (2015). https:// doi.org/10.2139/ssrn

91. Redi, J., Povoa, I.: Crowdsourcing for rating image aesthetic appeal: Better a paid or a volunteer crowd? In: Proceedings of the 2014 International ACM Workshop on Crowdsourcing for Multimedia, pp. 25–30. ACM (2014). https://doi.org/10.1145/2660114.2660118

92. Reid, G.B., Shingledecker, C.A., Eggemeier, F.T.: Application of conjoint measurement to workload scale development. In: Proceedings of the Human Factors Society Annual Meeting, vol. 25, pp. 522–526. Sage Publications Sage, Los Angeles (1981)

93. Reips, U.D.: Standards for internet-based experimenting. Experimental Psychol. **49**(4), 243 (2002)

94. Reiter, U., Brunnström, K., De Moor, K., Larabi, M.C., Pereira, M., Pinheiro, A., You, J., Zgank, A.: Factors influencing quality of experience. In: Quality of Experience, pp. 55–72. Springer, Cham (2014)

95. RICE, J., BAKKEN, D.G.: Estimating attribute level utilities from "design your own product" data—chapter 3. In: Sawtooth Software Conference (2006)

96. Rogstadius, J., Kostakos, V., Kittur, A., Smus, B., Laredo, J., Vukovic, M.: An assessment of intrinsic and extrinsic motivation on task performance in crowdsourcing markets. In: ICWSM (2011)

97. Ross, J., Irani, L., Silberman, M., Zaldivar, A., Tomlinson, B.: Who are the crowdworkers?: shifting demographics in mechanical turk. In: CHI 2010 Extended Abstracts on Human Factors in Computing Systems, pp. 2863–2872. ACM (2010)

98. Ryan, R.M., Connell, J.P.: Perceived locus of causality and internalization: examining reasons for acting in two domains. J. Personality Soc. Psychol. **57**(5), 749–761 (1989). https://doi. org/10.1037/0022-3514.57.5.749

99. Ryan, R.M., Deci, E.L.: Intrinsic and extrinsic motivations: classic definitions and new directions. Contemporary Educ. Psychol. **25**(1), 54–67 (2000). https://doi.org/10.1006/ceps.1999. 1020

100. Ryan, R.M., Deci, E.L.: Self-determination theory and the facilitation of intrinsic motivation, social development, and well-being. Am. Psychol. **55**(1), 68–78 (2000). https://doi.org/10. 1037/0003-066X.55.1.68

101. Sauro, J., Dumas, J.S.: Comparison of three one-question, post-task usability questionnaires. In: Proceedings of the SIGCHI Conference on Human Factors in Computing Systems, pp. 1599–1608. ACM (2009)
102. Shen, C., Zhao, Q.: Webpage saliency. In: European Conference on Computer Vision, pp. 33–46. Springer, Cham (2014)
103. Siddiqui, M.W., Eichelbaum, J., Llombart, V.C., Rosin, P.: Study project: predicting effort of microtask (2017)
104. Silberman, M.S.: What's fair? rational action and its residuals in an electronic market. Unpublished manuscript (2010). http://www.scribd.com/doc/86592724/Whats-Fair
105. Silberman, S., Milland, K., LaPlante, R., Ross, J., Irani, L.: Stop citing Ross et al. 2010, who are the crowdworkers? (2015)
106. Sparck Jones, K.: A statistical interpretation of term specificity and its application in retrieval. J. Documentation **28**(1), 11–21 (1972)
107. Stevenson, A.: Oxford dictionary of English (2010)
108. Tabachnick, B., Fidell, L.: Using Multivariate Statistics. Always Learning. Pearson Education (2012)
109. Tremblay, M.A., Blanchard, C.M., Taylor, S., Pelletier, L.G., Villeneuve, M.: Work extrinsic and intrinsic motivation scale: Its value for organizational psychology research. Canadian J. Behav. Science/Revue Canadienne des Sci. du Comport. **41**(4), 213–226 (2009). https://doi.org/10.1037/a0015167
110. Union, I.T.: ITU-t recommendation p. 800: Methods for subjective determination of transmission quality (1996)
111. Urošević, S., Milijić, N., et al.: Influence of demographic factors on employee satisfaction and motivation. Organizacija **45**(4), 174–182 (2012)
112. Walther, D., Koch, C.: Modeling attention to salient proto-objects. Neural Netw. **19**(9), 1395–1407 (2006)
113. Warr, P.: Work, Happiness and Unhappiness. Taylor & Francis, New York (2011)
114. Wechsung, I.: An Evaluation Framework for Multimodal Interaction: Determining Quality Aspects and Modality Choice. Springer, Cham (2014)
115. Wechsung, I., Weiß, B., Kühnel, C., Ehrenbrink, P., Möller, S.: Development and validation of the conversational agents scale (CAS). In: Interspeech, pp. 1106–1110 (2013)
116. Winkler, S.: On the properties of subjective ratings in video quality experiments. In: International Workshop on Quality of Multimedia Experience, QoMEx 2009, pp. 139–144. IEEE (2009)
117. Wolters, K.M., Engelbrecht, K.P., Gödde, F., Möller, S., Naumann, A., Schleicher, R.: Making it easier for older people to talk to smart homes: the effect of early help prompts. Univ. Access Inf. Soc. **9**(4), 311–325 (2010)
118. Zhao, Y., Zhu, Q.: Evaluation on crowdsourcing research: current status and future direction. Inf. Syst. Frontiers **16**(3), 417–434 (2014)
119. Zhao, Y.C., Zhu, Q.: Effects of extrinsic and intrinsic motivation on participation in crowdsourcing contest. Online Inf. Rev. **38**(7), 896 (2014)
120. Zijlstra, F.R.H.: Efficiency in work behaviour: A design approach for modern tools. Ph.D. thesis, TU Delft, Delft University of Technology (1993)
121. Zijlstra, F.R.H., Van Doorn, L.: The construction of a scale to measure perceived effort. University of Technology (1985)

Printed in the United States
By Bookmasters